FUNDAMENTALS

of

ROBOT

TECHNOLOGY

An Introduction to Industrial Robots,
Teleoperators and Robot Vehicles

D J Todd

A HALSTED PRESS BOOK

JOHN WILEY & SONS
New York

First published in 1986 by Kogan Page Ltd
120 Pentonville Road, London N1 9JN

Published in the U.S.A. by
Halsted Press, a division of
John Wiley & Sons, Inc., New York.

Library of Congress Cataloging-in-Publication Data
Todd, D.J.
    Fundamentals of robot technology.
    "A Halsted Press book."
    Bibliography: p.
    Includes index.
    1. Robots, Industrial. 2. Manipulators (Mechanism)
3. Automated guided vehicle systems. I. Title.
TS191.8.T63   1986   670.42'7   86-45039

ISBN 0-470-20301-3

Printed and bound in Great Britain

Contents

Chapter 1

Introduction

Robotics is a subject without sharp boundaries: at various points on its periphery it merges into fields such as artificial intelligence, automation and remote control, so it is hard to define it concisely. It is the branch of engineering whose subject is, obviously, robots, but there is no universal agreement on what constitutes a robot, although many definitions have been proposed, some of which are given later.

The boundaries of robotics are not only vague but shifting. Robots are evolving quickly, and our ideas with them, so that we expect more and more intelligence from machines. A machine which at one time is regarded as a robot may in a few years come to be thought too primitive or inflexible to merit the name. But if some machines are leaving the domain of robotics, others are entering, as it becomes possible to automate more tasks so that, for example, it becomes reasonable to envisage autonomous mobile robots travelling and working in the country unattended for long periods. Also, it may be argued that the boundaries of robotics are subject to changes of fashion: at the time of writing robots and so-called 'high-tech' devices generally are prominent in the news media and have value as commercial symbols, and so almost any piece of domestic hardware may be heralded as robotic by its advertisers.

Given this fluid situation it is unwise to insist on a rigid definition of 'robot' or 'robotics', but the following list of characteristics seems to be essential for a true robot.

1) A robot must be produced by manufacture rather than by biology. (This does not rule out the eventual use of artificial biochemically produced structures such as muscles.)
2) It must be able to move physical objects or be mobile itself. (This excludes simulations and control systems for static plant.)
3) It must be a power or force source or amplifier. (This excludes those teleoperated arms which merely replicate an operator's hand movements by a mechanical linkage. Nevertheless, teleoperators in general are a legitimate subject for robotics: a comprehensive treatment must sometimes stray beyond the bounds of a purist's definition.)
4) It must be capable of some sustained action without

intervention by an external agent.

5) It must be able to modify its behaviour in response to sensed properties of its environment, and therefore must be equipped with sensors.

A less formal view of a robot is that it is a machine possessing functional arms or legs, or else is a driverless vehicle. Other definitions emphasize intelligence, by which is meant the human-like ability to perform a variety of incompletely specified tasks involving perception and decision making.

Definitions of 'robot' and 'robotics'

The term 'robotics' was coined by Isaac Asimov in about 1940. Because of its origin in science fiction it is only slowly becoming a respectable word, and it is not found even in some good recent dictionaries. ('Roboticist' will probably take even longer.) Respectable or not, these words describe a coherent discipline and its practitioners, and will not go away. The origin of the word 'robot' around 1917 with Karel Capek is described in many books on robots; a brief but adequate account is given elsewhere (see bibliographic notes).

Other definitions in robotics

The term *manipulator* is used here to mean any device with an arm bearing a hand or gripper; thus it includes both industrial robots and telemanipulators. An *industrial robot* is a manipulator which automatically repeats a cycle of operations under program control. The identification of 'industrial robot' with programmed manipulator is unfortunate since other machines such as automatically guided vehicles (AGVs) are really also industrial robots. This identification may become weaker in time, but for now the term industrial robot must be assumed to refer to a programmed manipulator if no contrary indication is given. The official definitions of robot issued by the national robotics associations such as the British Robot Association (BRA) and the Japanese Industrial Robot Association (JIRA) are mostly of industrial robots in this sense. An example is that of the Robot Institute of America (RIA):

A robot is a reprogrammable and multifunctional manipulator, devised for the transport of materials, parts, tools or specialized systems, with varied and programmed movements, with the aim of carrying out varied tasks.

These organizations usually classify robots into four or more classes and sometimes into generations. There is little agreement about these classes, and in particular on whether pick and place machines and telemanipulators count as robots.

A *telemanipulator* is a manipulator whose actions are remotely controlled by a human operator, sometimes by mechanically replicating his hand movements and sometimes by obeying pushbuttons or joystick controls. Such a manipulator is often called a *teleoperator,* but in the view of Vertut and Coiffet (see bibliographic notes) a teleoperator, although it can be just a telemanipulator, more generally refers to a system of which a telemanipulator is merely a part, being moved about by some kind of transporter or vehicle.

An alternative term for teleoperator is *telechir,* coined, along with the associated subject name *telechirics,* by M.W. Thring as preferable because both halves are the same language: Greek for 'distant' and 'hand'. (He also invented the term 'sceptrology', meaning the technology of mechanical aids for the disabled.)

For completeness some terms are defined which, while not relevant to robotics as a practical subject at present, tend to be associated with robots, particularly in fiction and speculation about the future.

An *android* is an, as yet imaginary, robot of human appearance and physical abilities. There is no agreement on whether an android must be built from engineering materials or grown in some biochemical way.

A *cyborg* is a being part machine and part biological. One would not wish to argue that a person with artificial hip joints or heart valves is a cyborg, which raises the question of how much has to be mechanically replaced before a person counts as one. 'Cyborg' is a hybrid of 'cybernetics' and 'organism'. Cybernetics is the science of control systems in engineering and biology; the word was invented by Norbert Wiener.

Connections between robotics and some related subjects

ARTIFICIAL INTELLIGENCE

Artificial intelligence (AI) is dealt with in Chapter 11; for now it is sufficient to make one or two general remarks. First, the boundaries of AI, like those of robotics, are rather fluid, particularly where AI merges into psychology and the other sciences of mind and brain in nature. Indeed, robotics has been regarded by some as a branch of AI, but equally AI could be said to be a subset of robotics, if robotics is interpreted liberally.

From a scientific or philosophical point of view the most interesting area of the AI–robotics interaction lies in the possibilities for making robots which are more like those of science fiction, i.e. mobile intelligent autonomous agents. This is touched on in Chapter 11. In terms of the practical robotics of today and the immediate future, however, the relevance of AI is mainly that it provides, or promises to provide, a number of useful techniques for enhancing performance. The general theme of these is making robots more intelligent, in a down to earth sense, by incorporating adaptability, sensing, problem solving and so on. There is also an opposite connection: robots for AI instead of AI for robotics; robots can be useful tools for developing AI techniques.

FLEXIBLE MANUFACTURING SYSTEMS, FACTORY AUTOMATION, COMPUTER-AIDED MANUFACTURING

There are often good reasons for installing an isolated industrial robot, but robots can also be regarded as elements in a larger system encompassing an entire manufacturing process. There is a whole range of options, from a *manufacturing cell,* in which a group of machine tools and robots makes a component or assembly, to attempts to automate almost an entire factory. As the system gets larger, the emphasis shifts from the individual machining or assembly process to issues of communication between the machines and the coordination of the whole enterprise. The aim is to gain increased control over the manufacturing done in a factory, so that the work is distributed among the machines efficiently, it is possible to make many varieties at once while keeping track of individual assemblies, and so on. It may even be possible to link the factory to a computer-aided design (CAD) process in such a way that the machine tool settings are derived automatically from the design.

This subject is beyond the scope of this book, but it may be useful to explain a few terms as they are commonly used in connection with robots.

A *flexible manufacturing system* (FMS) is a coordinated set of machine tools and their loading devices (often robots), which can produce a range of items. It implies that there is a communications network associated with the machines so that they can be programmed, while the system is running, with the settings for each item to be made.

Factory automation means the linking together of most of the operations of a factory so that the whole process is under automatic control; sometimes the idea of an unmanned factory is spoken of, but in practice there are almost always some operations still done by workers (in addition to the maintenance staff).

The terms *computer-aided manufacturing* (CAM), *computer-integrated manufacturing* (CIM) and *computer-aided engineering* (CAE) generally refer to flexible manufacturing and factory automation together with issues such as stock control and the linking of CAD with manufacturing, but without any universally agreed definitions.

Bibliographic notes

Since some references are cited in several chapters, they are given in a list at the end of the book rather than being cited in full in the bibliographic notes for each chapter.

A brief account of the origin of the word 'robot' and of Capek's work is given in Reichardt (1978). This also gives several references to the history of toy robots and automata.

An interesting if unorthodox approach to the philosophy of robots is contained in Thring (1983).

Chapter 2

Geometric Configurations for Robots

Introduction

Robots take a bewildering variety of forms: arms of all shapes, vehicles with all possible arrangements of wheels or legs, and devices which although clearly robotic are neither vehicles nor arms. This chapter makes sense of this variety by explaining how the functions required of a robot can be met by combinations of mechanical elements such as links and joints. It concentrates on manipulation robots.

In treating the geometric or spatial aspects of robot design, we start from the proposition that a robot is a machine for moving things around. The thing it moves may be a workpiece, a tool, a passenger or a cargo: in general, a payload. The robot may have to move two or more payloads in a co-ordinated fashion. The payload is usually an object with definite boundaries, but this is not universally true: it may be part of something extended, such as a membrane or rope. It is often rigid, but some payloads are flexible, or partially liquid, or even active things such as live animals.

The movement of the payload(s) must be *relative* to something: often to the ground, but another case is when one item must be moved relative to a second, their absolute position being of no consequence. An example is the movement of a tool relative to a workpiece. It is possible to keep the workpiece fixed and to move the tool, to keep the tool fixed and to move the workpiece, or to give some dimensions of movement to the workpiece and the rest to the tool.

The distinction between arms and vehicles

If we wish to move an object about in space, relative to something fixed such as the ground, six dimensions are needed to specify its situation: three for its position and three for its orientation or attitude. These must be controlled in some way, by connecting the object to the ground with a link allowing the transmission of the forces and torques needed to support and move the object. If the link is continuous and fixed to the ground we have a manipulator; if it is made by a self-contained device which propels itself about on a surface or through a medium we have a vehicle. It is important to note that there is no sharp boundary between vehicles and manipulators:

as a vehicle becomes more constrained by rails and by its propulsion method it becomes indistinguishable from part of a manipulator. Whether a particular device is a vehicle or not is sometimes just a question of point of view. For example, there are some computer plotters whose pen is moved by a small steerable wheeled carriage connected to the computer only by a ribbon cable: is this a vehicle? Another marginal case is that of the arm mounted on a carriage running on a short track: is the carriage a vehicle, or merely another joint of the manipulator?

Nevertheless, for most purposes the distinction is clear enough, and in the case of free-ranging vehicles carrying a manipulator it is usually possible to regard the two functions as separate: the vehicle is thought of as a mobile platform for the, separately controlled, arm. However, in the future, control methods will be needed which allow good coordinated movement of arm and vehicle.

Structural elements of manipulators*

As remarked previously, an industrial robot or teleoperator must control the position and orientation of an object using a mechanical link to the fixed world. This link could in principle take an unusual form, such as a collection of inflatable bags or a set of flexible elements such as bimetallic strips. It need not even involve physical contact: in some circumstances the object could be suspended by magnetic or other forces.

However, nearly all robots actually use rigid links connected by rotating or sliding joints, with the occasional use of tension elements such as wires, tapes and cables. This is true even of the apparently flexible elephant-trunk-like arms which have sometimes been made: in this case the links are short and the joints many.

Degrees of freedom and number of joints

A machine made of rigid links connected by joints is characterized by its number of degrees of freedom. A joint can have more than one degree of freedom (e.g. a ball joint allows rotation about three independent axes) or it may not contribute any degrees of freedom at all. This happens if two or more joints are coupled so that they can only move together.

We are usually interested in the position and attitude of the payload at the end of the linkage chain. From this point of view some degrees of freedom may be redundant, i.e. there is more than one way of accomplishing a given movement of the payload. This happens if two joints can rotate in the same plane (i.e. their axes are parallel), or if two translational joints are not at right angles. In the simplest case the machine has as many

---

* 'Manipulator' as used here includes both programmmable industrial robots and teleoperators.

degrees of freedom as joints, whose number is in turn equal to that of the controllable degrees of freedom of the payload. This number is at most six, three for position and three for orientation. Practical robots often have fewer, especially in orientation since many tasks can be designed so that orientation is either unimportant or held constant. Occasionally they have more than six.

Types of joint

The basic kinds of motion possible at a joint (or articulation as it is sometimes known) are rotation and translation or sliding. A single-axis rotary joint such as a hinge is called a *revolute* joint; a joint with a single direction of sliding, and with no rotation, is called *prismatic*. Other possible joints are the cylindrical joint, allowing both sliding and rotation, the helical or screw joint, the spherical or ball joint, and the flat planar joint in which one half is constrained to slide in a plane. For analysical purposes these can usually be regarded as combinations of revolute and prismatic joints.

A notation for joints uses the symbols R for revolute and P for prismatic. If two or more joints coincide their symbols are joined by a brace: thus a ball joint is denoted by RRR, and a cylindrical joint by RP. This notation allows a compact description of an arm as a chain of joint symbols starting from the base of the machine. The notation is not suitable for parallel structures in which two points are connected by more than one chain of links.

Of the joints found in robots, only revolute and prismatic joints are powered. Ball joints do occur, for the attachment of push rods and in parallel linkages, but these are not powered.

Since rotary and rectilinear (straight-line) motion can be interconverted by a rack and pinion, a cable and pulley, or a screw, a revolute joint can be powered by a rectilinear actuator, or a prismatic joint by a rotary one. Nevertheless, it is natural to match rotary actuators with revolute joints and rectilinear with prismatic. So prismatic joints are often powered by hydraulic or penumatic cylinders, and revolute joints by electric or hydraulic motors. All the other possibilities such as pneumatic rotary acutators can also be found.

Revolute joints are generally to be preferred to prismatic. A prismatic joint although conceptually as simple as a revolute one is in practice a more complex mechanism, or more difficult to manufacture. (This partly reflects the ready availability of rotary bearings, which is an aspect of the prevalence of rotary machinery in general.) Further, a prismatic joint takes up a lot of space, as shown by Figure 2.1 which compares a prismatic joint with an angular one giving the same displacement.

Prismatic joints can be compounded as in a telescope, but at the cost of added complexity. A final argument for the revolute joint is that angular

17

position sensors are more readily available than rectilinear ones. (This may be only a minor point if the prismatic joint is driven by a rotary actuator, provided that the gearing introduces no errors.)

However, a prismatic joint can be made very rigid; and it generates rectilinear motion without coordinated servo joint control.

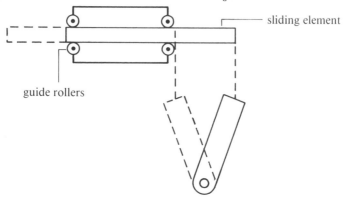

Figure 2.1 *Comparison of prismatic and revolute joints, showing that for a given displacement a prismatic joint takes up more space.*

Construction of joints

Revolute joints simply use conventional bearings. The possibilities for a prismatic joint are more various. Some examples are shown in Figure 2.2. Various designs of rectilinear ball bearing, some with recirculating balls, also exist. The machine slide (Figure 2.2 (c)) is not very suitable for robots because its high friction discourages rapid movement.

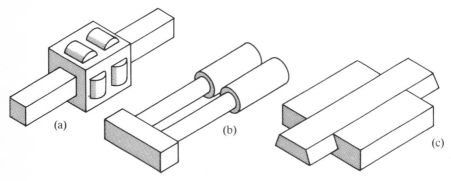

Figure 2.2 *Construction of prismatic joints: (a) roller guides, (b) pair of hydraulic cylinders, (c) machine slides.*

Parallel linkages

Most manipulators take the form of an arm consisting of a chain of links connected end to end, or *serially*. However, it is also possible to achieve an equivalent motion by connecting the links side by side, in parallel. A two-dimensional illustration is given in Figure 2.3.

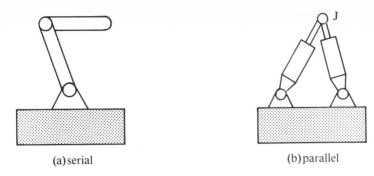

(a) serial   (b) parallel

Figure 2.3 *Comparison of serial and parallel linkages.*

In Figure 2.3(b), if both rams extend equally, the joint J will move vertically. If the rams move by unequal amounts J can be made to move in any direction, including horizontally. The advantage of a parallel linkage is increased rigidity due to the triangulated structure. The most notable configuration of this kind allows the control of all six degrees of freedom of the payload using six prismatic actuators. It was invented for use in flight simulators and has been used in at least one experimental industrial robot, the GEC Gadfly assembly robot (Figure 2.4).

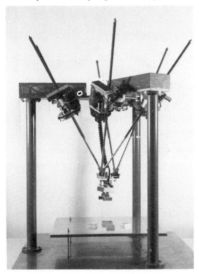

Figure 2.4 *A robot with a parallel linkage, the GEC Gadfly (courtesy GEC Research Ltd).*

This structure has the advantages of rigidity, of low inertia, and of uniformity, since all the joints, with their motors and sensors, are identical. A disadvantage is its limited range of movement, particularly in rotation. The Gadfly was designed for printed circuit board component insertion, for which this limitation is not serious. Its six prismatic joints use leadscrews driven by servomotors. The gripper can move at speeds of the order of a metre a second, and the positioning repeatability is around 0.1mm.

Combinations of serial and parallel linkages are possible. For example, it may be advantageous to fit a serial wrist, giving large angular ranges, to a parallel arm of high rigidity.

Constrained linkages

In another variation on the theme of linkages, two or more joints are constrained to move similarly. This connection is usually an extra link or a belt and its most usual purpose is to keep two parts of the machine parallel so that, for example, the axis of the gripper is always horizontal. The principle is shown in Figure 2.5.

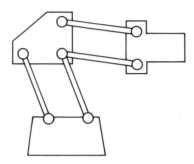

Figure 2.5 *Constrained linkages: a pair of parallelogram linkages keeps the wrist horizontal at all times.*

Distributed manipulators

Before concentrating on arm-like manipulators, we must amplify the earlier remark that some motions can be given to the workpiece and others to the tool. In the simplest form of this division the workpiece is mounted on a set of slides like those of a milling machine, producing translational movement in the two horizontal dimensions and possibly the third, vertical dimension, while the tool is suspended above on a mounting which swivels about whatever axes of rotation are required. It may also slide vertically. Such a design is capable of great rigidity and therefore positional accuracy. It would be useful for precision grinding of a complex shape where there are large forces between the workpiece and the tool, tending to deflect the tool's path.

Robot transporters and workpiece positioners

These may be regarded as conferring extra degrees of freedom on a robot. Figure 7.28(f) shows how a transporter can increase the working volume of a robot so that it can serve several machines. A transporter can be a carriage on a fixed track as in this example, or can be a vehicle. Transporters are particularly important for teleoperators, as explained in Chapter 8 (in particular see Figure 8.8).

A workpiece positioner (known as a manipulator by some manufacturers, such as ESAB) allows the orientation, and sometimes the position of the workpiece, to be controlled while the robot applies a tool such as a welding gun to it. Some positioners resemble lathes, with a headstock and tailstock between which a very large workpiece can be clamped; others have a workpiece table which can rotate about one or more axes.

When equipped with a positioner a robot in effect becomes a distributed manipulator with many degrees of freedom. Positioners have several uses:

1) Generating an accurate circular motion by rotating the workpiece; this is used in arc welding of circular seams.
2) Turning a workpiece over so that the robot can get at both sides.
3) Helping with loading and unloading; the positioner can be double ended so that it can present one workpiece to the robot while a new workpiece is being loaded into the other end.
4) Performing a similar function to a transporter, by effectively increasing the working volume of the robot.

Arm configurations

In most manipulators there is a clear distinction between the function of the arm itself, which for the purpose of this section does not include the wrist or the gripper, and that of the wrist. The function of the arm is to position the payload, and that of the wrist is to orientate it. So an idealized manipulator has long links in the arm, to allow large displacements, and links of zero length in the wrist. (In other words, the axes of the three revolute joints of the wrist intersect in a point.) This section describes some common arm configurations. They all have three joints such that the tip of the arm can move in three dimensions (Figure 2.6).

CARTESIAN

The Cartesian or $x, y, z,$ arrangement is the only one to use just prismatic joints, corresponding to the dimensions of the Cartesian coordinate system. This is the mathematically simplest system as far as translational movements are concerned. It is easy to calculate what joint movements are needed to move the payload from one place to another, and arm movement does not affect payload orientation. This is advantageous when dealing with a world dominated by right angle geometry; an example is inserting components into printed circuit boards. Robots for this purpose often hang down from a gantry rather than standing like a pillar. Since the slideways for the two motions are supported at both ends, a gantry machine is easier to make rigid, and is the most rigid of the common structures for industrial robots. Two examples of Cartesian robots are shown in Figure 2.7.

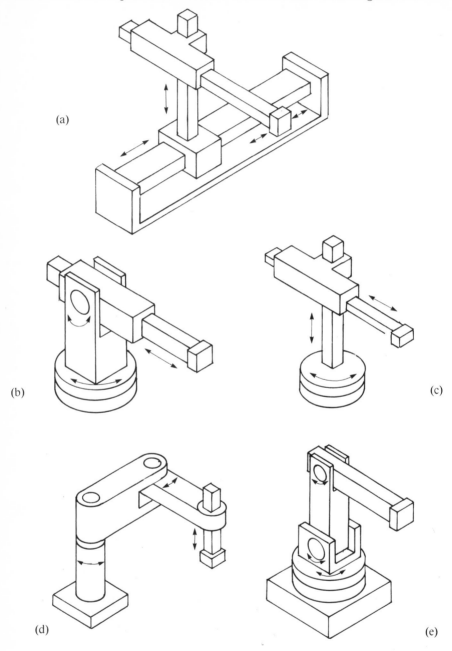

Figure 2.6 *Five common arm geometries: (a) Cartesian, (b) polar, (c) cylindrical, (d) SCARA, (e) jointed or anthropomorphic.*

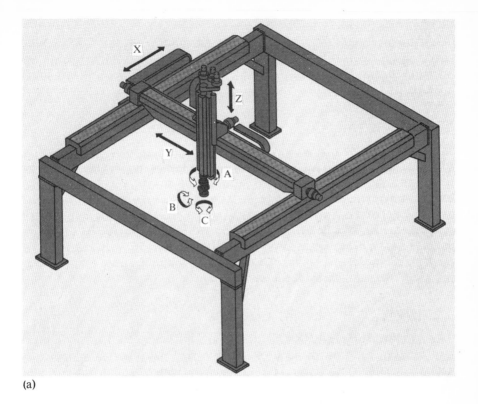

(a)

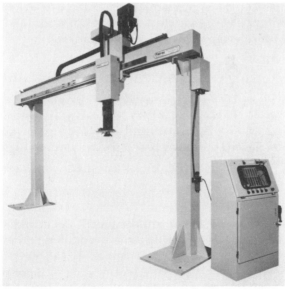

(b)

Figure 2.7 *Two Cartesian robots with a gantry strucure: (a) KUKA IR 400, (b) Fairey gantry robot.*

POLAR

The spherical polar or r, θ, ϕ configuration (Figure 2.6 (b)) was adopted in the Unimate, the first industrial robot. This design is used mostly for machine loading, being well suited to a long straight reach into a press or moulding machine.

CYLINDRICAL

This cylindrical or r, z, θ configuration is found mostly in pick and place arms and robots for parts feeding. As in the Cartesian geometry the wrist is automatically kept in a constant attitude (apart from rotation about the vertical axis), and so it is suitable for tasks such as the assembly of an electric motor where the assembly is dominated by a vertical axis along which components such as bearings and shafts are to be inserted. Its main advantage over the Cartesian arrangement is that the robot can be surrounded by the machines it serves and can swing right round to cover a large working area. It is shown in Figure 2.6(c).

HORIZONTALLY JOINTED ARMS

The horizontally jointed arm is shown in Figure 2.6(d). This configuration is often referred to as a 'SCARA' arm, for selective compliance assembly robot. This means that the vertical axis at the gripper is kept rigidly vertical, and the vertical prismatic joint can apply a force for insertion during assembly while the horizontal motion is allowed to be somewhat compliant to take up small positioning errors. The SCARA arm is often used for assembly, like cylindrical and Cartesian arms, and for the same reasons. It is more compact than a cylindrical robot.

VERTICALLY JOINTED OR ANTHROPOMORPHIC ARMS

The vertically jointed or anthropomorphic arm (Figure 2.6(e)), sometimes said to use revolute coordinates, is very popular. It is compact (and so has a large working volume for its size) and avoids prismatic joints. It can be set up in several normal operating positions, as shown in Figure 2.8. Some examples are shown in Chapter 7.

OTHER ARM DESIGNS

Other joint combinations are sometimes found. An example, which may be described as PRRPRR in the notation introduced earlier, is shown in Figure 2.9. Suspended over the workpiece, it can exert a high force downwards and is used for pressing and drilling of large workpieces.

Extra joints can be inserted into any of the previously described arms, in order to reach round corners or to provide motion in a particularly useful

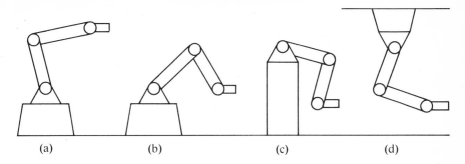

(a)   (b)   (c)   (d)

Figure 2.8 *Mounting positions for a vertically jointed arm.*

direction without relying on the coordinated movement of several joints. The whole arm can be mounted on an extra source of motion. An example is a track parallel to an assembly line conveyor, so that a welding robot can be moved along at the same speed as the car bodies it is working on.

A more radical change is the attempt to produce more flexible arms by stacking many identical joints in series. An example is shown in Figure 2.10.

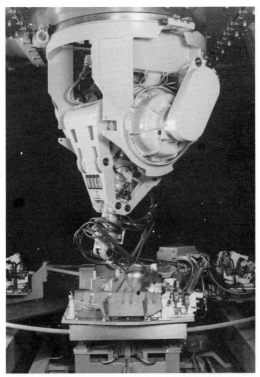

Figure 2.9 *An unusual arm geometry: the KUKA IR 260/500, for jobs needing a large pressing force (courtesy KUKA Welding Systems & Robots Ltd).*

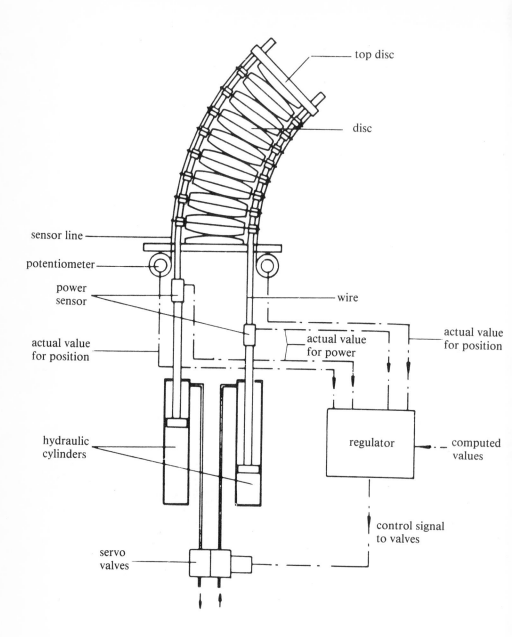

Figure 2.10 *The mechanism of the Spine robot: four cables pulled by hydraulic rams pass along the edges of a stack of discs (each disc is mounted in a square frame with a cable hole at each corner). Differential pull in a pair of cables causes the stack to bend as shown. It can bend in two planes at once, so each stack has two degrees of freedom. The complete robot contains two such stacks and a three axis wrist.*

Tension structures

Before moving on to wrist and end effectors, a brief note will be included on an alternative approach to manipulator construction. An object can be positioned (and orientated) by cables, usually pulled by winches. Figure 2.11(a) shows how an object confined to a plane can be positioned by two active cables together with a means of keeping them in tension; this could be the object's weight or a third cable. The confinement to the plane might be by gravity or by a linkage of rigid members. Such a robot does exist. Still in two dimensions, if extra cables are added the orientation of the object can also be controlled.

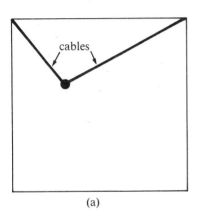
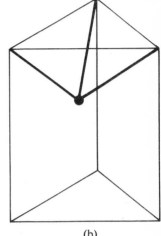

(a)   (b)

Figure 2.11 *Tension structures: (a) two dimensional, (b) three dimensional; in each case the arrow indicates that weight or another source of tension is needed to keep the suspension cables taut.*

In three dimensions an object can be positioned by three cables kept in tension, as in Figure 2.11(b); this system is used to suspend a television camera over a stadium, forming what is in effect a giant teleoperated robot. With six cables the geometry is like that of the Gadfly described earlier, and attitude about three axes can be controlled as well. (This again assumes that gravity or an extra cable keeps the controlling cables in tension.)

Finally, note that tension elements in the form of cables, chains or tapes are often used in transmitting mechanical power to the joints of rigid-link manipulators; they correspond to the tendons of animal limbs.

Wrists

The mathematically ideal wrist allows rotation of the held object about three axes at right angles, such as those shown in Figure 2.12, although wrists with fewer axes are also common. There are several systems of names for the three axes; the system adopted here is that of pitch, roll and yaw.

27

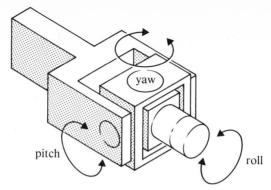

Figure 2.12 *The basic three-axis wrist.*

Before going further, a point worth noting about mechanisms with three rotational axes (and this applies to gyroscopes and arms as well as to wrists) is that things tend to go awry at large angles: it is possible for two of the axes to become aligned, or nearly so, a condition known as degeneracy. For example, in Figure 2.12 if the hand were to yaw through 90° from the straight ahead position the roll joint would coincide with the pitch joint. This would make it impossible to rotate the payload about the long axis of the arm. The design of wrist shown here can actually yaw through only a limited angle, perhaps ± 45°, so degeneracy is not a severe problem, but some wrists can yaw through 90° or more.

It would be mathematically tidy, and often useful in practice, if the axes of rotation passed through some specified point in the payload, such as the tip of a tool or the centre of a workpiece. This is not usually possible; the joints are almost always some distance behind this point. A more realistic objective is to make the three joint axes intersect at a point somewhere in the wrist, and this is commonly done.

It is difficult to design a satisfactory three-axis wrist. The main problem is this: if the joint actuators are attached directly to the joints, the wrist will be heavy and bulky, but if they are not then some mechanical transmission must be devised to couple the wrist joints to actuators mounted on the arm.

A wrist must meet some or all of the following requirements:

1) The mass at the end of the arm should be minimized. Otherwise the arm will have to be stronger than it need be, and because of excessive inertia the robot will be slow.
2) The volume of the wrist mechanism should be minimized as it is liable to intrude into the work space and to collide with obstacles.
3) The wrist must allow transmission of power to the gripper. This can be done by a mechanical linkage, which tends to be complicated, or the wrist may be bypassed by a flexible hose or

cable. Cables or optical fibre bundles from sensors in the gripper may also have to be accommodated.

4) Each joint must have a useful angular range. There are two reasons why this can be more difficult than it appears at first sight. First, if a gimbal-like arrangement is used the movement of the innermost element will be restricted by the gimbal ring; secondly, any tendons, cables or hoses to the end effector have limited flexibility.

5) It is sometimes desirable to allow a compliant or force-sensing device to be incorporated.

The main approaches to wrist design will now be described.

DIRECT DRIVE TO EACH AXIS

The actuators are usually hydraulic vane rotary actuators, geared hydraulic motors or electric servomotors. Figure 2.13 shows a robot with a hydraulic motor on each of the three wrist axes.

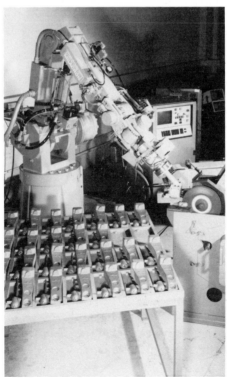

Figure 2.13 *A three-axis wrist with direct drive to each axis by a hydraulic actuator. The robot, a Cincinnati Milacron T³–586, is cutting off unwanted parts of a casting by holding it against an abrasive wheel (courtesy of Cincinnati Milacron).*

GEAR DRIVES

Mechanical engineers have shown great ingenuity in devising gear trains for driving three-axis wrists. The most usual elements are bevel gears and concentric shafts. A common approach is to design a two-axis joint and to mount this on a relatively simple third joint. Figure 2.14 shows two variants of a two-axis joint, one allowing roll and pitch and the second allowing pitch and yaw. In Figure 2.14(a), if bevel gear R is driven one way while P is driven the other way at the same speed, there will be no pitch motion but shaft S (the roll axis) will rotate. If both R and P are driven at the same rate in the same direction the cage K, and therefore the whole hand, will rotate about the pitch shaft, with no roll. Such wrists are found on some five-axis educational robots.

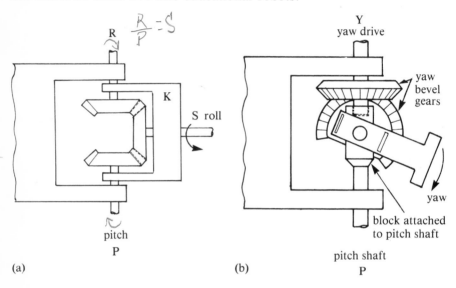

Figure 2.14 *Two-axis wrist joints: (a) pitch and roll, (b) pitch and yaw; the shafts P, R and Y are often driven by bevel gears and shafts (as shown in Figure 2.16), or by cables and pulleys.*

The device of Figure 2.14(b) works similarly, in pitch and yaw instead of pitch and roll.

Some of these two-axis joints (such as that shown in Figure 2.14(b)) require that to rotate them in one axis alone both input shafts must be driven at the same rate. This can be done by controlling the speed of the two motors separately, or by using a differential gear as shown in Figure 2.15. The differential adds the angular velocity of the pitch and yaw motors in such a way that driving the wrist in pitch alone does not affect the yaw angle: the pitch angular velocity is passed to both shafts on the pitch axis, one directly and the other via the differential.

Figure 2.16 shows how a two-axis (pitch and yaw) joint can be mounted on a third joint whose axis lies along the forearm to make a three-axis joint.

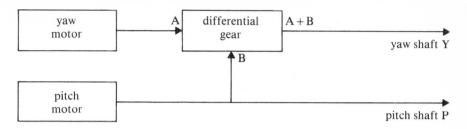

Figure 2.15 *Use of a differential gear to decouple yaw and pitch for the wrist shown in Figure 2.14(b). If the yaw motor alone is powered, only the yaw shaft rotates. If the pitch motor alone is powered, not only does it drive the pitch shaft but it also drives the yaw shaft at the same rate. This is so there is no relative movement between the yaw shaft and the yaw bearing block, so the wrist is not driven in yaw.*

In Figure 2.16(a) the whole forearm rotates, which has the advantage that the pitch and yaw drives need no decoupling from the roll axis. In Figure 2.16(b) such coupling does occur and must be compensated for either by servo control or by differentials.

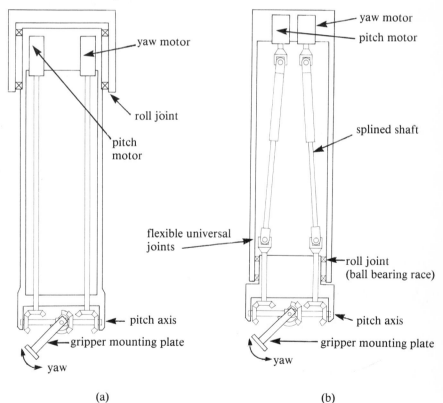

(a)   (b)

Figure 2.16 *Two simple designs for a three-axis wrist.*

Even more complex gear trains are possible. For example, the three-axis wrists can be modified to allow the transmission of power to a gripper by fitting an extra set of bevel gears on shafts running through the pitch and yaw shafts; the principle, for a two-axis pitch/roll wrist, is shown in Figure 2.17.

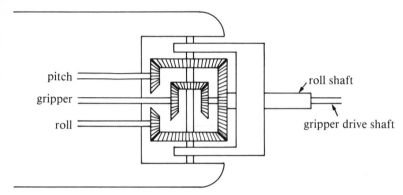

Figure 2.17 *Transmission of gripper drive through a two-axis wrist using bevel gears and concentric shafts.*

A very neat wrist mechanism invented by Cincinnati Milacron can now be found on many robots (Figure 2.18); it can be recognized by its spherical or biconical housing. Motion supplied by three concentric input shafts from motors in the arm is converted into three-axis motion which, for some orientations at least, is equivalent to roll, pitch and yaw, in a way impossible to describe except by a three dimensional model.

COMPLIANCE IN WRISTS

During assembly, if a shaft or bolt is to be inserted into a closely fitting hole, or a collar slid onto a shaft, it may jam if there is the slightest alignment error. The problem is usually explained in terms of the peg and hole shown in Figure 2.19. It is assumed that the chamfered edge of the hole will guide the tip of the peg in if there is a slight positional error. The problem is then that the peg may well jam at an angle in the hole, and the only way to free it is to rotate the peg towards the axis of the hole, or to move the top sideways towards this axis.

A rotation of this kind can be accomplished by a linkage called a remote centre compliance (RCC) linkage, shown in Figure 2.20. It has the property that any sideways movement by the wrist while the peg is in contact with the hole is compensated for by the linkage in such a way that the peg rotates about a point near its tip (a centre remote from the linkage, hence its name) in the right direction.

Other passive linkages have been devised. The alternative is to instrument the wrist with force sensors, usually strain gauges, and to use servo control to drive it in the right direction. An active system is more complex and slower than a passive one but can in principle cope with a wider range of jamming and misalignment problems.

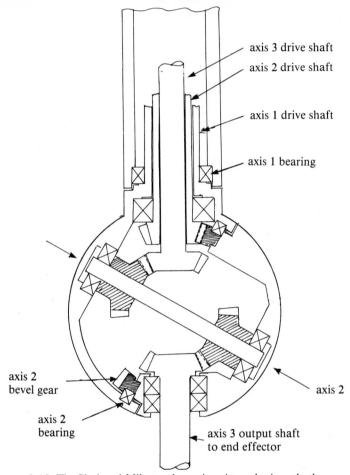

Figure 2.18 *The Cincinnati-Milacron three-axis wrist mechanism: the three axes are not at right angles. A gripper drive could be passed through the assembly by the addition of shafts running through the centre of those shown, and connected by bevel gears.*

End effectors (grippers, tools and hands)

The purpose of an arm, whether an industrial robot, a telemanipulator or a prosthesis, is to carry a tool or workpiece. The most general way of accomplishing this is to provide a hand which, like the human hand, is almost infinitely adaptable. Such hands are still in the research stage, as described later. Industrial robots generally have a two-jaw gripper, a special purpose gripper, a permanently mounted tool or a fitting for interchangeable tools.

TWO-JAW GRIPPERS

The crudest kind of two-jaw gripper is shown in Figure 2.21(a). It is rarely used except in very cheap robots owing to its tendency to displace the load

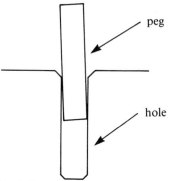

Figure 2.19 *An example of alignment error — a peg being inserted into a hole.*

sideways as the jaws close. The version of Figure 2.21(b) is preferred. The jaws are coupled by gears or a linkage. The variant of Figure 2.21(c) has been fitted with jaw faces matching a particular part. Those of Figures 2.21(d), (e) and (f) are suitable for gripping objects with parallel faces.

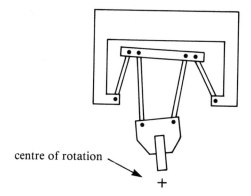

Figure 2.20 *A remote centre compliance (RCC) linkage, shown in two dimensions. In practice the linkage is three dimensional, allowing compliant motion in two planes, and occupies a cylindrical housing.*

Such grippers can use any kind of power source, but peneumatic actuation is particularly suitable because of its inherent springiness: if the workpiece starts to slip, some resilience in the jaws may enable the gripper to keep its hold. However, grippers are also available with jaws powered by servomotors which allow servo control of jaw separation or gripping force. The gripping force may be measured by strain gauges or the motor current.

Gripper jaws can be lined with resilient and high friction materials, but it is hard to find such materials with a long life.

SPECIAL PURPOSE MECHANICAL GRIPPERS

Grippers have been designed to fit all sorts of objects. A gripper for car wheels is shown in Figure 2.22 and another which can handle a complete car door and which includes the tools for bolting the door onto the body is illustrated in Figure 2.23. Some more examples are shown in Figure 2.24. Most grippers grip by friction or mechanical fitting. Other methods are possible, such as adhesive tape or spikes (for soft materials).

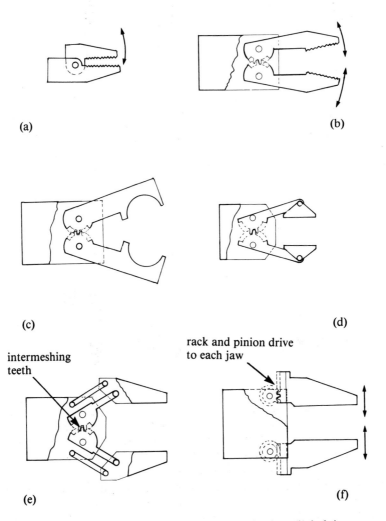

Figure 2.21 *Varieties of a two-jaw gripper: (a) single moving jaw; (b) both jaws move symmetrically, coupled by gears; (c) jaws shaped to match a cylindrical workpiece; (d) self adjusting parallel jaws; (e) standard design of parallel jaw gripper; and (f) sliding jaws driven by rack and pinion for straight-line motion.*

35

Figure 2.22 *A special gripper for holding a car wheel (courtesy of KUKA Welding Systems & Robots Ltd).*

Figure 2.23 *A special gripper for mounting a car door. It incorporates two nut runners for attaching the door to the body shell (courtesy of KUKA Welding Systems & Robots Ltd).*

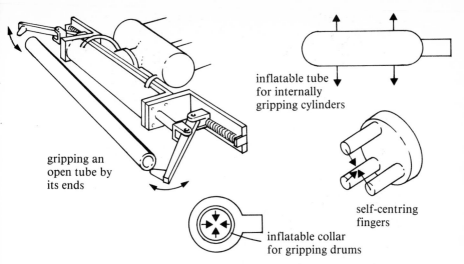

inflatable tube
for internally
gripping cylinders

gripping an
open tube by
its ends

self-centring
fingers

inflatable collar
for gripping drums

Figure 2.24 *Special purpose grippers*

VACUUM AND MAGNETIC GRIPPERS

Suction cups are used in egg packing machinery and glass handling as well as in robotics. A common way of producing the (partial) vacuum is to supply compressed air to a venturi; this is simple and has a low initial cost but is noisy and consumes power. The alternative is a vacuum pump; this is quieter and produces a better vacuum. Suction, with a water venturi, also works under water.

It is easy enough to grip a smooth flat object by suction, but harder if the surface is rough and the shape complex. Adapting grippers have been designed with many small suction cups in parallel, mounted by ball joints to individual cylinders (Figure 2.25) or springs. An extension of this idea is to apply a vacuum to a porous gripper surface, which works like many tiny suction cups in parallel; such a gripper can pick up objects with flat but rough surfaces, such as bricks.

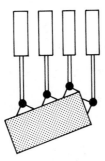

Figure 2.25 *Adapting vacuum gripper with suckers on individual pneumatic or hydraulic cylinders. It is lowered with the pistons free to move until it matches the workpiece, then the pistons are locked in place by closing valves.*

Magnetic grippers can use permanent magnets or electromagnets. An electromagnet can, of course, easily be turned off, but when this is done it is desirable to apply a pulse of reversed current. This cancels any magnetization induced in the workpiece and ensures a quick release. If a permanent magnet is used there must be a mechanical device for detaching the workpiece, or else the robot can pull free once the workpiece is secured at its destination. Permanent magnets can be designed to produce shallow magnetic penetration so that only the top layer of a stack of steel sheets is picked up.

Both magnetic and vacuum grippers are of limited load capacity. Another property they share is that if an object is held by a flat surface it will tend to slide if it is at an angle to the horizontal or if it is accelerated too rapidly. Vacuum and magnetic grippers also need a continuous supply of power.

TOOLS

Chapter 7 describes several applications such as arc and spot welding, spraying and grinding for which the robot carries a tool. Other tools are gas torches, ladles, drills, wrenches, screwdrivers and inspection equipment.

In deciding whether a tool is suitable for robot use several factors must be taken into account:

1) the weight of the tool;
2) the positional and angular accuracy with which it must aligned with the workpiece;
3) any sensing needed to use it;
4) the rigidity with which it must be held (tools such as grinders and wrenches exert a large reaction on the robot);
5) cables, hoses and other supplies;
6) reliability (a tool such as a screwdriver which jams once in a thousand operations might be acceptable by a person but is useless for a robot).

TOOL AND GRIPPER CHANGING

Tools are generally attached to the outermost link of the wrist, in place of a gripper. The attachment may be permanent (but allowing the tool to be changed by maintenance staff) or may use some quick-release mechanism which the robot can operate. Tools and grippers can then be kept in a magazine from which the robot can select one. A variation of this is for the gripper to be fixed but to have interchangeable jaws. If there is any danger that the tools might be put in the magazine in the wrong order, an identity mark such as a bar code can be put on each tool, to be read by a sensor in the wrist interface. The tool interface may have to allow connection of electric, pneumatic or hydraulic supplies. Figure 2.26 shows a KUKA robot with automatic tool changing.

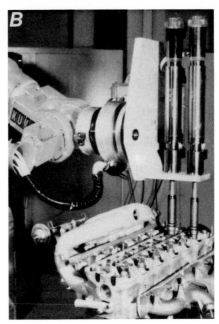

Figure 2.26 *A robot with a quick-change tool adaptor. In (a) a gripper is attached, and in (b) a double bolt driver (courtesy of KUKA Welding Systems & Robots Ltd).*

ANTHROPOMORPHIC AND OTHER ADAPTIVE HANDS

An anthropomorphic hand is one resembling the human hand in having a thumb opposed to several fingers, the thumb and fingers each having several joints. It is a special case of the adaptive hand which by using several joints can wrap round or otherwise grip a wide range of objects. We must also distinguish between teleoperated hands in which each finger joint follows that of a human operator, who provides all the intelligence needed for adaptive grasping, and computer-controlled hands. Satisfactory operation of a computer-controlled hand is difficult because, in addition to the problem of actuating such a complex mechanism, it must be possible for the computer to sense joint angles, contacts and pressures, and to interpret these in terms of an internal model of the held object and the grasping process. This is a long way from being accomplished except in special cases, and no industrial robot is today fitted with an adaptive hand of any complexity.

The adaptability discussed so far is an active capacity needing intelligence, but the term is also used for structures which conform passively to a variety of shapes. This can be done by elastic jaw liners, inflatable grippers or other means.

Two-dimensional adaptive grippers

Many objects can be grasped well enough by a hand with two digits swinging in the same plane, like a two-jaw gripper except that the digits are more complicated. The simplest advance on the two-jaw gripper is to incorporate extra joints (Figure 2.27). Such hands have rarely been used, perhaps because the added complexity confers little advantage. An extension of this idea, again with limited practical success, is shown in Figure 2.28 in which a whole chain of finger segments, collectively actuated by a single grip wire and a release wire, can wrap round an object of arbitrary shape.

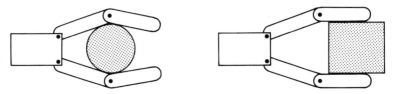

Figure 2.27 *Adapting gripper with two jointed fingers; the joints may all be independently powered or some may be coupled together.*

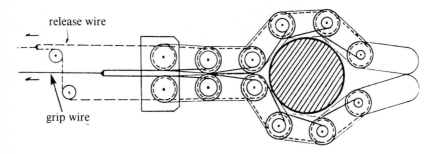

Figure 2.28 *Adaptive gripper with many joints but needing only two actuators.*

Anthropomorphic hands

The most advanced hand at present is the dextrous hand (DH) being developed by the University of Utah and MIT. (A similar hand is also being developed at Stanford.) It has three fingers and a thumb, each with four joints, and a three-axis wrist (Figure 2.29).

Each joint is powered by a pair of pneumatic single-acting cylinders via a pair of tendons; a simplified example is shown in Figure 2.30. The tendons are polymer composite tapes, 3 mm wide, using Kevlar tension fibres. The actuators use a glass cylinder and a loosely fitting graphite piston for low friction. This design needs 38 actuators, which are stacked in a dense array in the forearm. Each actuator is controlled by a specially designed electropneumatic servo valve designed for servo control of pressure rather than flow. The whole system is designed for a very fast response and so the hand will be roughly as fast as a human one.

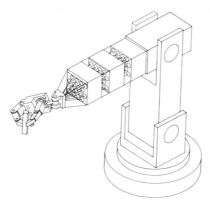

Figure 2.29 *The Utah/MIT Dextrous Hand (after Jacobsen et al. 1985). The 38 pneumatic cylinders which actuate the 19 joints of the hand are arranged in three stacks, with up to 16 in each stack (this allows a maximum of 48 actuators). The tendons for the rear, or proximal, and middle stacks pass through the spaces between the front, or distal, cylinders. The wrist actuation method is not shown; also, the three axis arm on which the hand is mounted is only schematic; it is not part of the design.*

Although the mechanical difficulties of making a good anthropomorphic hand are extreme, they constitute only part of the problem: there is a massive control problem in doing any of the many high precision manipulations possible with the human hand. Some research has been done with simpler hands, usually with three digits, and quite a lot is known about the possible interactions, with and without friction, between a held object and three digits. (Three is a significant number because it allows the hand to cup an object in a cage of fingers without relying on friction.) Devices like the Utah/MIT hand will allow such research to be less constrained by poor mechanics.

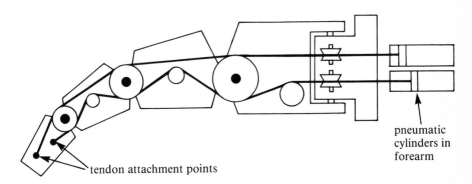

pneumatic
cylinders in
forearm

tendon attachment points

Figure 2.30 *Tendon arrangement for the distal joint of one finger of the Utah/MIT Dextrous Hand.*

Bibliographic notes

For further information on snake-like structures see Lhote *et al.* (1984), Thring (1983), Hirose and Umetani (1977), Hirose, Umetani and Oda (1983), and Taylor (1983).

Lhote also describes several complex gearing schemes for three-axis wrists, active and passive compliance mechanisms, and gripper designs. Several chapters of Brady *et al.* (1983) are devoted to the theory of compliance.

The Cincinnati Milacron three-roll wrist is described in Stackhouse (1979).

Many special-purpose grippers, and details of the practical aspects of magnetic and vacuum grippers, are described in Engelberger (1980). A large collection of illustrations of hands and grippers is presented in Kato (1982).

The Utah/MIT hand is described in Jacobsen *et al.* (1985). Some examples of adaptive gripping by three-fingered hands are given in Brady *et al.* (1983) and, in a somewhat mathematical treatment, in Kobayashi (1985).

A recent textbook on the theory and practice of hands and manipulation (including extensive material on the Stanford/JPL hand) is Mason and Salisbury (1985).

Chapter 3

Operation, Programming and Control of Industrial Robots

Types of industrial robot and their methods of operation

An industrial robot, unlike a telemanipulator, is driven through a sequence of movements by a program of some kind. The program is executed by a controller; the basic relationship between the controller and the robot is shown in Figure 3.1. The controller turns on the joint actuators (throughout this chapter the terms 'joint' and 'axis' are used interchangeably) at the appropriate times, while signals from the joint sensors are returned to the controller and used for feedback. The types of controller, methods of programming and details of joint servo control are discussed in the following sections. We begin with the classification of industrial robots.

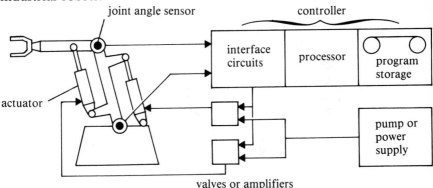

Figure 3.1 *The basic architecture of an industrial robot.*

Industrial robots can be classified by the method of control and by the method of teaching or programming; although certain control methods and teaching methods are almost always used together, in principle the two bases of classification are separate.

The main classes of control are as follows:

1) pick and place (non-servo),
2) point to point,
3) continuous path.

It is important to note that these classes are a reflection of the kinds of hardware and control law used to drive the joints. Unfortunately certain terms, such as point to point, are used both in this context and in that of teaching. As will be seen, a robot which is continuous path from the point of view of control theory may be taught in a point to point way.

PICK AND PLACE MANIPULATORS

Pick and place or *limited sequence* manipulators, which are not always counted as robots at all, use mechanical stops to set two stopping positions on each axis. The joint must travel backwards and forwards between these two end stops, whose positions can be adjusted when the machine is set up; it is not possible to select any intermediate stopping point, although sometimes extra stops can be inserted for particular parts of the program. This is usually done by solenoid-operated pins which when extended prevent the arm moving beyond them.

Pick and place robots are usually pneumatic since there is no need for servo control of position; the speed of each axis is limitied by an orifice on the actuator. They are fast and relatively cheap; as the name suggests they are limited to the transfer of parts from one place to another. The parts are always well orientated at the pick-up site, and often only three or four degrees of freedom are needed. Some designs are modular, allowing the user to select how many degrees of freedom to provide. An example of a pick and place robot is shown in Figure 3.2.

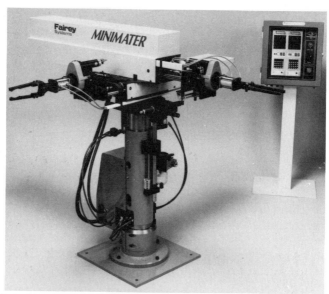

Figure 3.2 *A pick and place or limited sequence robot. It is pneumatically powered and has two grippers so it can do the two stages of a transfer process on two successive workpieces at once (courtesy of Fairey Automation Ltd).*

POINT TO POINT ROBOTS

Point to point robots have servo position control of each axis and can go through a sequence of specified points. The path between these points is unspecified. There can be any number of stopping positions in each axis. The program for such a robot consists of a series of points; for each point all the joint angles (or distances in the case of prismatic joints) must be specified.

This method of operation implies that the actuator for each joint is controlled by a position servo (see the section 'Servo control of actuators' in this chapter). Speed* and acceleration are not controlled explicitly, although in practice they are limited in some way, if only because the actuators have a finite top speed. The result is that each axis of the arm moves at a nearly constant speed between programmed points. Such an arm is clearly unsuitable for tasks requiring speed control. It is really just a pick and place machine with an arbitrarily large number of programmable positions.

A further limitation is that it is not in general possible to guarantee that a segment of the programmed path will take a desirable form. For example, the payload cannot be made to move in a straight line unless this coincides with the axis of a prismatic joint. This is why assembly robots so often have a prismatic joint parallel to the main axis of the assembly (usually vertical).

CONTINUOUS PATH ROBOTS

Continuous path robots do not go through a finite list of target points but can, ideally, execute a smooth path of any shape, with continuous variation of speed as the arm moves along the path. This requires not only servo control of the velocity of each joint but that several joints move at once in a coordinated way, whereas for a point to point robot it is possible, although not compulsory, to move only one joint at a time.

Methods of teaching and programming

How does an industrial robot determine what movement to make next? There are two extreme possibilities: the movement is calculated at the time, or it is replayed from an existing program or recording. The first method is necessary if the robot is to respond continuously to sensory inputs, e.g. if it is to follow a surface using a proximity sensor. Otherwise, the second method can be used. This is the method in general use. An intermediate case is for the program to have branches selected by sensor signals, or to accept certain values, such as a desired gripper rotation angle, from an external source.

---

* 'Speed' is used here to mean the linear velocity of a prismatic joint and the angular velocity of a revolute one, and, when applied to the whole arm, to mean roughly the linear velocity of the payload. 'Position' and 'acceleration' are used in a similarly flexible way.

This section concentrates on non-sensory robots. Such a robot is always driven by a program, whether this is a sequence of indivisible actions, a sequence of target positions or a continuous record of position or velocity. If this program is acquired by somehow causing the robot to go through the required motions while these are recorded, the robot is said to be taught. Whether teaching is a form of programming or an alternative to it is merely a question of definition on which there is no universal agreement. This section describes four basic methods of teaching and programming. Many robots allow a choice or combination of these methods so that taught sequences can be interspersed with or embedded in a program which has been written off-line. Also, when the user teaches a robot a pair of path end points, the detailed trajectory of the arm between them may be generated by the same software regardless of whether the end points were taught by leading through or were programmed off-line.

PROGRAMMING PICK AND PLACE ROBOTS

The crudest form of programming is the setting up of a pick and place machine. This has two parts: the mechanical end stops are set in place for each axis, and the sequence in which the joints operate is programmed. The sequencer on early machines was a mechanical device such as a multiposition rotary switch with several cams on its spindle, each cam operating contacts to switch power to the solenoid valves for the pneumatic cylinders. Each end stop had a limit switch to detect when the joint reached the end of its travel. The activation of a limit switch caused the sequencer to rotate a few degrees to its next position, at which the connections were made to turn on the next joint. The order in which the joints operated was determined by the pattern of connections between the sequencer and the solenoid valves. This pattern was made on a plugboard, and putting the plugs in the right places was how the manipulator was programmed.

In more modern pick and place machines the mechanical sequencer and plugboard are replaced by an electronic, microprocessor-based controller, allowing the arm to be programmed by typing a series of axis identification codes on a small keyboard. The controller has interfaces for reading the limit switches and driving the valves. It can be a general purpose industrial controller as used for machine tools. Such controllers allow the storage of several programs, and these may have subroutines and allow a wide choice of time delays between actions. One controller may be able to handle several robots in a coordinated way, and to interlock their operation with that of other machinery.

WALK-THROUGH TEACHING OR PENDANT TEACHING*

This is the most usual method with point to point servo robots. A hand-held box or 'pendant' has buttons, toggle switches or joysticks corresponding to each axis of the arm, which cause the axis to be driven under power (but possibly faster or slower than it will move when the program is played back). The user drives the robot to a required position using these controls and then presses a button which causes all the joint position sensors to be read and their values stored; the robot is then driven to the next position on its required path and so on.

This method of teaching has certain consequences not obvious at first sight. An important one is that since the path between two programmed points is unspecified, and since there will usually be several joints active at once, the arm may not approach a target point from the same direction as it did during teaching. Therefore an extra point is often inserted into a program so that the approach to a critical point requires movement of just one axis. Intermediate points are also inserted to take the path round obstacles.

A related consequence is that, if the movement between two programmed points uses two or more joints, then it is likely that one joint will have a shorter distance to travel and will therefore finish before the others. The resulting trajectory of the payload will therefore consist of a series of arcs with abrupt direction changes. Both of these effects can be avoided, at the cost of reduced overall speed, by building the program exclusively from segments in which only one joint is moved at a time.

WALK-THROUGH TEACHING WITH PATH CONTROL

Some of the problems just mentioned can be overcome provided that continuous path control is available, together with suitable software. When the user has entered two points on the desired path, the robot's computer calculates a straight line between the points which the robot can follow, at a speed chosen by the user, at playback time.

This method can be used only with a continuous path robot, although to the user teaching the robot it resembles teaching a point to point robot. An example of its uses is in paint spraying in cases where there are long straight runs of the spray gun, which can be specified by teaching just the start point and end point of the run.

Robots with this path control feature usually have a lot of computing power and good servo control and so can offer other facilities, such as generating a circular path by interpolation given three points on its circumference, or the ability to move in a straight line in some useful coordinate system.

---

* 'Walk-through' is used here to mean teaching with a hand-held pendant as explained below, and 'lead-through' to mean physically leading the arm, as described shortly. As often as not in existing sources, however, these terms are used in the opposite sense: lead-through for teaching with a pendant and walk-through for physically leading an arm. It would perhaps be best to abandon these terms and use 'pendant teaching' and 'physical arm leading' instead.

LEAD-THROUGH TEACHING OR PHYSICAL ARM LEADING

In this form of teaching the user carries out the required motions with his own hand, while holding some device for recording the path taken. This device may be the manipulator itself or a replica arm, the 'master arm' or 'teaching arm', which is geometrically similar to the robot but is light enough to move easily, is unpowered and has angular or displacement sensors on its joints similar to those on the robot. The signals from these are recorded and become the program which the robot plays back. The program can be replayed at a fraction or multiple of the speed at which it was recorded.

Since the joint positions are recorded continuously, this method can be used for continuous path robots, and is commonly used for tasks such as some kind of paint spraying in which the movements are complex and continuous. A point to point robot can also be taught by lead-through, the joint positions being recorded just at those moments when the user presses a button.

OFF-LINE PROGRAMMING

The alternative to teaching a robot by driving it through its cycle of operations is to type in a program at a computer terminal. In the simplest case the program consists of a series of commands of the form 'move axis A through distance D'. These commands are expressed in some language designed for robot programming. Since the program which actually controls the robot is not in this form but is instead concerned with primitive operations such as turning valves on and off, the program that the user enters must be compiled to yield a control program in the computer's machine code. A later section discusses some of the languages which have been designed for programming industrial robots.

THE IMPLICATIONS OF SENSING FOR ROBOT CONTROL

Whatever the method of teaching or programming, provision must be made to incorporate sensor tests or measurements, if only interlock signals, into the program. Teaching pendants allow the user to insert an instruction causing the arm to await an external signal between arm movements. This is easy to arrange since the controller only has to do one thing at a time. A more complex case is when it must monitor some sensor input while the robot is moving and take appropriate action if an input occurs. More complex still is when sensor readings, such as the speed of a conveyor, must be continuously used, in real time, to compute the trajectory of the arm as it moves along the conveyor.

Types of controller and program memory

As explained earlier, pick and place arms use microprocessor-based controllers. The program consists of a sequence of axis movements, each being specified by an identification code. Other elements in the program are time delays, entered as a number of seconds or milliseconds, and instructions to wait for interlock signals.

Point to point robots need a means of storing joint positions; so, for example, a six-axis robot each of whose axes had a positional accuracy of one part in 1024 (2^{12}) would need to store six 12-bit words for each programmed point on the path. A capacity of several hundred points is usual. Many forms of digital memory have been used, such as magnetic bubble devices, cassettes tape, floppy discs, and semiconductor memories. Since the robot must not forget its program when switched off, some form of non-volatile memory is needed — hence the predominance of magnetic methods. Often there is a memory for the current program and also removable storage using cassettes or floppy discs so that programs can be stored away from the robot and transferred between robots.

The memory of a continuous path robot is similar, except that many more data values must be stored. In the simplest case the robot functions essentially as a point to point robot with many closely spaced points, i.e. the program is recorded by sampling the joint angles 50 or 60 times a second. When played back all these points are retraced, and so there may be thousands of them if the program takes more than a few seconds. In this system each joint is driven by a position servo, but the points are so close together that it does not come to rest at each position but is in continuous motion.

The computing power of robot controllers varies enormously, from a single 8-bit microcomputer for the whole robot to systems having a 16-bit single-board computer per joint, in which case the issue of coordinating all these computers becomes significant.

Analysis and control

Analysis means finding a mathematical description of a robot in relation to its surroundings, which will allow the calculation of the geometric and dynamical quantities used in controlling it. Control in the sense of this and the following sections means operating the robot's actuators so as to produce a specified path and velocity of the payload.

Both analysis and control are complicated by the multiple-link nature of manipulators, which implies that both geometric calculations, such as computing the distance from the end effector to a fixed workpiece given all the joint angles, and dynamical ones, such as finding the effective moment of inertia acting on a particular actuator, can be difficult. The subject is necessarily highly mathematical, and its details are of interest to a small proportion of those who work with robots: therefore this book aims only to give an overview of the issues involved.

It is nearly always assumed that a robot can be regarded as a chain of rigid links connected by revolute or prismatic joints at which actuators, regarded as torque or force generators, act. The control of flexible structures is in its infancy and will not be discussed.

With this assumption, there is a set of important problems in analysis and control, and most of the literature on robot control addresses one or other of these. Some have accepted solutions; others are the subject of research. They are as follows:

1) formulating the kinematic equations (joint coordinates to world coordinates);
2) solving the kinematic equations (world coordinates to joint coordinates);
3) the forward problem of dynamics — finding the motions resulting from joint torques;*
4) the inverse problem of dynamics — finding the torques needed to produce a given motion;
5) specifying a trajectory between target points on the path;
6) actuator servo control — for a single actuator, how to drive it so as to produce a specified position, velocity or torque.

Any task for a robot involves some combination of these. In the simplest case, that of a manually taught point to point robot, only (6) is needed, whereas if the tip of the arm of a continuous path robot is to describe a circle specified in world coordinates most of them are needed. (World coordinates are those in which the base of the robot is fixed.)

These six problems are now described.

FORMULATING THE KINEMATIC EQUATIONS

Kinematics is concerned with distances and angles and translational and angular velocities and accelerations, but not with forces, masses, torques and moments of inertia, which are the province of dynamics.

If the position and orientation of, say, a tool held by a robot is required relative to the fixed world and if all the link dimensions and joint angles are known, the calculation can be made in a standard way. A three-dimensional coordinate system is embedded in each link, as shown in Figure 3.3. For each joint a transformation is found between the coordinate system of the two links it connects; if this operation is applied to each joint in succession, the relationship between any two links, including the payload and the fixed base of the robot, can be found. These transformations are

---

*To avoid writing 'force or torque', 'velocity or angular velocity' and so on, this discussion uses terms such as 'torque' with the understanding that the equivalent translational or rotational term is implicit.

expressed as equations called the *kinematic equations* of the manipulator. It is convenient to use *homogeneous coordinates* for the system of each link, and in this case the kinematic equations take the form of matrix multiplications. Therein lies their advantage, for a chain of transformations is simply a chain of matrix multiplications and so the relationship between any two links is easily expressed. (In homogeneous coordinates a fourth element w is appended to the usual three coordinates x, y, and z. It represents a scale factor. Its introduction allows translations, rotations and scalings all to be accomplished by matrix multiplication, and it is the usual formalism adopted in computer graphics. It has the advantage for robots that revolute and prismatic joints can be treated in the same way.)

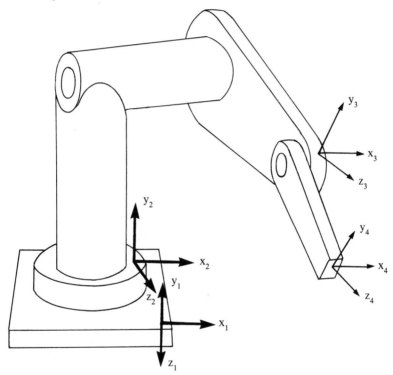

Figure 3.3 *Coordinate systems attached to the segments of an arm.*

SOLVING THE KINEMATIC EQUATIONS

The previous problem is straightforward in principle, although when written out in full the transformation matrices can be complicated, but the inverse problem may have a less systematic solution, and not always a unique one. This problem is to take a given position and orientation of the gripper or payload and to compute from it the joint angles of the arm. It is important since tasks are often specified in terms of the position of the payload and we want the robot system to work out for itself what joint angles are needed to generate this payload state. The methods of solving it, in effect solving the kinematic equations, are not amenable to a simple explanation, and indeed are still a subject of active research. The problem is discussed in several books on manipulator analysis and control (see bibliographic notes).

The Jacobian formulation

The kinematic equations deal in coordinates (angles and distances) measured from the base axes of each coordinate set. However, it is often convenient to deal in differential changes, i.e. for a joint or a chain of joints we compute the small movement expressed in one coordinate system equivalent to a small movement expressed in the other. This gives a way of calculating how velocities and accelerations transform between coordinate systems. The relationship between differential changes in two coordinate systems is expressed by a matrix called the *Jacobian*, and control methods based on its use are referred to as Jacobian control. An example of the use of the Jacobian is in calculating the joint velocities needed to produce a given payload velocity.

THE 'FORWARD' AND 'INVERSE' PROBLEMS OF DYNAMICS

Going beyond kinematics we meet the problem of the relationship between the torques applied at the joints by the actuators and the resulting motions of the arm. As with kinematics there is a pair of problems, one being the inverse of the other. The first of these is the problem of how a system of links moves when torques are applied to its joints; this is the 'forward problem'. It amounts to solving the equations of motion of the system. Once again the methods of doing this, which are based on either the Lagrangian or the Newton–Euler formulation of dynamics, are too complex to be described here.

The 'inverse problem', which fortunately is somewhat easier to solve, is, given a stipulated motion of the system, to calculate the required joint torques. It is clearly relevant to the control of robots. It can be thought of as formulating the equations of motion in such a way that the torques are given by functions of angles and velocities.

SPECIFYING TRAJECTORIES

The previous four problems are ones of providing some basic mathematical tools which can be used for many purposes. The fifth is the question of how to calculate the theoretical path we want the robot to follow. Note that we may speak (i) of the trajectory (the terms 'path' and 'trajectory' are here used interchangeably) of an individual joint, meaning the angle as a function of time for that joint considered in isolation, (ii) of the trajectory of the whole arm, meaning the set of all joint angle functions and (iii) of the trajectory of the end effector or arm tip, which is a path through three-dimensional space.

The issue of trajectory generation is only applicable to robots with continuous path control. It does not arise if the robot simply reproduces a path recorded by the walk-through method, but does arise in several other situations. First, we may wish it to travel a straight line in three-dimensional space between two points which have been programmed in, either off-line or by leading through. The most obvious way of generating a path in this case is to use a linear interpolation to create a string of closely spaced points along the desired line. A more difficult but related task, which is particularly useful for welding circular seams such as the joint between the end of a pipe and a flange, is to draw a circle. The circle may be specified by giving three or four points on its periphery by leading through, from which the rest of the circle can be calculated, or, if specified by off-line programming, its centre, radius and orientation in space will be given, and in either case a set of closely spaced points along its circumference can be calculated. The circle is thus approximated by a many-sided polygon along which the robot-held tool will travel. The error introduced by this approximation, for a circle 10 cm in diameter and for points calculated every 2 mm along the path, would be 0.01 mm, which is negligible in nearly all cases.

A third example of trajectory calculation is when some external motion, often the movement of a conveyor carrying the workpiece past the robot, must be compensated for. It is convenient to teach the robot, by lead-through or walk-through, to spray or weld a stationary workpiece; then during production the workpieces, such as car bodies, move past the robot at a speed which may or may not be constant. Provided that it can be measured, this speed can be used to recalculate the path of the tool so that it tracks the moving workpiece while carrying out the programmed task. If the conveyor speed is constant this calculation can be done in advance; otherwise the robot controller must compute the next points on the trajectory in real time.

So far, trajectories have been discussed mainly in terms of the path of the end effector in three-dimensional space, but as remarked earlier a trajectory includes the orientation of the payload as well, and one of the important aspects of trajectory calculation is keeping the payload in the correct attitude while it translates in space. A common requirement is to keep the orientation constant; another is to keep a tool facing a fixed point, so it rotates about an imaginary centre. This is the sort of motion the human hand makes when using a spanner.

What does a trajectory, of any shape, consist of? It is actually a program for a continuous path robot, i.e. a series of closely spaced points, the position of all joints being specified for each point. To give a concrete example, the points calculated on the periphery of a circle might be spaced every 2 mm, which for an end effector speed of 10 cms^{-1} corresponds to 50 points a second. Between any two of these points intermediate points are calculated by linear interpolation, and so the arm actually describes a series of straight lines 2 mm long. The number of intermediate points must be sufficient to present a virtually continuous input to the joint servos. By 'virtually continuous' is meant a sampling rate as high as that of the servo, if it is a digital one; this might be 300 samples per second, so the intermediate linearly interpolated points would need to be generated for every 1/300 s, or 1/3 mm apart.

SERVO CONTROL OF ACTUATORS

When a robot is moving freely (and not exerting a force as it would do in, for example, grinding) there are usually certain programmed points that it is required to move through, and it is the positional accuracy at these points which is the most important variable to be controlled. Therefore the joints are usually operated as position servos which can track a moving set point. Velocity is controlled by continuously varying the position input to the servo.

The theory of servomechanisms has been dealt with in a multitude of books; thus this section is limited to a suggestion of what is involved. The examples given use analogue servo systems for clarity, but modern servo systems are often digital, so the physical integrators and other components are replaced by parts of a program.

The form of a servo system depends on the nature of the actuator, which is a transducer of some input signal such as current or voltage to an output quantity such as position or torque. Further, an actuator may have 'invisible' servo systems built into it so that, for example, it appears as a proportional transducer of input voltage to output shaft angle but internally uses a motor which is fundamentally a transducer of current to torque. For the present purpose the principles can be illustrated by a motor which generates a torque proportional to its input voltage.

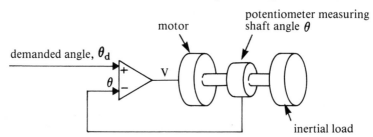

Figure 3.4 *A proportional angle servo system for an inertial load.*

Figure 3.4 shows the simplest proportional angular position servo, for a purely inertial load. The potentiometer measures the shaft angle θ; if there is an error between this and the demanded angle θ_d a voltage $V = k_p (\theta_d - \theta)$ will be applied to the motor, which will generate a torque T

$= Gk_p (\theta_d - \theta)$ and accelerate the load until the angle error is zero. (In these equations k_p is the constant of proportionality between the angle and voltage; G is the torque produced by the motor per unit input voltage.) Such a simple system is unstable unless there is some damping, i.e. a retarding torque proportional to speed. Damping can be added either by differentiating the position signal or by providing a tachometer to measure rotational speed directly. If the servo is to track a sustained rate of change of the demanded position, i.e. to work as a velocity servo as well as a position one, the velocity feedback signal can be made proportional to the velocity error rather than to the absolute velocity; such a servo is shown in Figure 3.5.

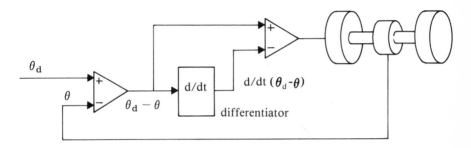

Figure 3.5 *A proportional-differential (PD) servo system for an inertial load.*

If the load is not purely inertial but is sustained, such as that due to gravity, then this system will show an error in its equilibrium state, since the only way it can maintain a steady torque is for there to be a steady position error. Therefore an extra feedback loop can be added which integrates the position signal. This has the effect that any sustained position error feeds back a signal which builds up the torque until the position error returns to zero. Such a servo is shown in Figure 3.6. It is called a PID (for position–integral–differential) servo.

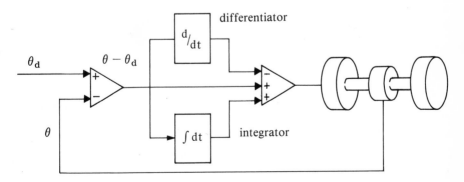

Figure 3.6 *A proportional-integral-differential (PID) servo system for an inertial load.*

Servos of this general nature, although often more complicated, are suitable for use with electric servomotors. Analogous servo systems can be designed for stepper motors or hydraulic actuators. Real robot systems have to take into account several additional factors, such as variable moments of inertia due to changing loads and arm configurations, non-rigidity of the structure, and the use of a robot as a force servo. For some tasks such as grinding it is necessary for the robot to follow a path while exerting a specified force. Another case is 'active compliance'. This means that the arm is controlled as a position servo in some directions while constrained by external forces in others. An example is inserting a peg into a hole, in which the robot acts as a position servo along the whole axis but is unable to move the peg sideways because of its contact with the hole.

Programming languages for industrial robots

As explained earlier, most robots can be programmed in some language which is compiled (or interpreted in the case of some slow robots intended for educational purposes) to yield the machine code which drives the robot. Many manufacturers provide a language for their own robots; meanwhile attempts are being made to develop universal robot languages, or to add robot-control features or subroutine libraries to languages such as Pascal or C.

The main concern of this section is to explain the capabilities desirable in a robot programming language. A first requirement is that it should provide an adequate range of the facilities expected of any programming language, such as flow of control (i.e. features such as BEGIN . . . END, IF . . . THEN . . . ELSE, and DO loops), arithmetical operations, data types, subroutines and functions. As well as these language features there should be an adequate range of support tools (editors, compilers, debugging facilities, file handling). Also, since robot control is an instance of real-time computer control, the language and the operating system, if any, under which it runs must be suitable for real-time control; this usually implies the ability to deal with interrupts and perhaps to allow several processes to run simultaneously while communicating with each other.

In addition to these general requirements, a language for robot control should have some of the following capabilities.

1) Geometric and kinematic calculations: functions and data types to allow the compact and efficient expression of coordinate systems in homogeneous coordinates and their transforms (matrix arithmetic); perhaps even very high level operations such as solving the kinematic equations.
2) World modelling: the ability to define objects by, for example, the space enclosed by a set of intersecting surfaces, to manipulate such objects as a whole, to 'attach' objects

(including parts of the robot) so that the system knows that if one object moves then any attached objects also move, and to test for collisions. Other functions can be imagined. World modelling is related to simulation and computer-aided design.

3) Motion specification: the ability to do the things described in the section on specifying trajectories. This implies functions such as linear interpolation, finding the equation of a circle from points on it, fitting curves through a series of points and so on. It may also be useful to specify sawtooth weaving, as used in arc welding, speed and acceleration, and the direction of approach to a certain point (important for assembly).

4) The use of sensing for program branching and servo control.

5) Teaching: the ability to accept path points taught by leading or walking through. This is perhaps an aspect of trajectory generation.

6) Communication with other machines.

7) Vision and other complex sensing (such as tactile imaging). Although it is more usual for such processing to be done in a separate system which feeds a simple result such as object orientation to the robot control system, there may be a place for these capabilities within the language itself.

Most existing languages have only a limited set of these capabilities.

Note that the detailed servo control of the actuators is usually not regarded as part of robot programming but is done by lower level software, probably written in a different language; this is often assembler language since speed is at a premium.

Dozens of robot languages have been developed. The earliest was MIT's language MHI in 1960. Its main robot-specific constructs are moves and sensor tests. A more general purpose language was WAVE, developed at Stanford in the early 1970s. It introduced the description of positions by Cartesian coordinates, coordinated joint motions and compliance by letting certain joints move freely under external loads. An influential language, which is still being extended, is AL. This provides Cartesian specification of motions, compliance, the data types and control structures of an Algol-like language, support for world modelling (such as attachment) and the concurrent execution of processes. A more recent language is RAPT, based on the machine tool language APT, and developed at Edinburgh University.

Robot manufacturers often provide a language to go with their products. A well-known example is Unimation's VAL for the PUMA robot.

Bibliographic notes

A standard textbook on manipulator analysis and control is Paul (1981); another well-known source is Brady *et al.* (1983), which is a book of collected papers. Other texts on analysis and control are Coiffet (1983), Vukobratovic and Potkonjak (1982), and Vukobratovic and Stokic (1982). Books on this subject are almost invariably heavy going except for the mathematically inclined. Chapter 1 of Brady can be recommended, however, as relatively accessible. It makes use of an imaginary two-joint manipulator to illustrate the mathematics of kinematics, dynamics, trajectories, servo control and so on.

Programming languages for robots are surveyed by Lozano-Perez (1982) and Gruver *et al.* (1983).

Chapter 4

Actuators for Robots

This chapter is primarily about actuators for manipulators, although the same considerations apply to the legs of walking robots. It does not cover the driving of wheels, since wheeled vehicle engineering is a mature subject and there is no shortage of books on traction motors, internal combustion engines and wheeled vehicle transmissions.

An actuator is any device for producing mechanical movement. It is usually considered to produce it from some non-mechanical power source, although there is a class of mechanical transmissions with clutches and related devices which are arguably actuators and can be regarded as such from a control point of view. By 'mechanical transmission' is meant the movement of solid components such as gears, linkages and cables; the movement of fluids, which might be regarded as mechanical in some contexts, is excluded.

With these assumptions an actuator can be thought of as a transducer of some non-mechanical form of power into mechanical power. This point of view leads to a classification of actuators along three dimensions:

1) the input power source,
2) the geometrical type of movement (linear, rotational etc.),
3) the dynamical type of output (whether the actuator is primarily a generator of position, velocity or force).

The assignment of a particular device to one of these categories may not be obvious because gearing is often built in to transform a rotary motion to a linear one or vice versa and because, as mentioned in Chapter 3, an actuator may have a servo system built in which changes the variable it apparently controls from, say, torque to angle. In general, actuators are at root sources of force or torque: any position or velocity output must ultimately arise from the acceleration which results when the actuator exerts a force.

This chapter follows the usual practice of taking the input power source classes as the major headings; within each of these the most common combinations of geometrical type and dynamical type will be described. Other combinations are possible, and are probably in use somewhere.

There are three types of power source in common use: pneumatic, hydraulic and electric.

Pneumatic actuation

Pneumatic actuators as used for industrial robots work on compressed air at a pressure of, typically, 10 bar (150 psi), which is provided as a standard service in many factories, so the robot does not need its own compressor. They are confined almost entirely to pick and place manipulators since the compressibility of air makes it difficult to design servo systems. In a pick and place machine the valves are either fully on or fully off, and each actuator stops only at the end of its travel. A pneumatic actuator is also often used for the gripper of an electric or hydraulic robot, where its elasticity is useful as it automatically limits the force which can be applied and can cope with variations in the size of the workpieces. Also, a pneumatic gripper actuator is very light and needs to be connected only by a narrow flexible tube which is easy to feed through a complex wrist. Pneumatic auxiliary devices, particularly jigs and clamps, are often used with an industrial robot and operated by the robot controller. Another pneumatic device, although not exactly an actuator, is a cylinder which balances the force of gravity on part of a manipulator, e.g. a vertical prismatic joint. This enables a smaller motor to be used. The pneumatic compensator is just an actuator kept at constant pressure by a valve so that it exerts a constant force regardless of joint position.

The most common type of pneumatic actuator is the cylinder or ram. These usually have a piston rod as shown in Figure 4.1, but rodless cylinders also exist in which a lug on the piston passes through a slot running the length of the cylinder. The slot is sealed by a flap which is pushed aside by the lug — a system first used on Brunel's atmospheric railway (with limited success, said to be because of rats eating the sealing flap which was made of greased leather). Rodless cylinders have the advantage of compactness. Another type of rodless cylinder uses a smooth cord instead of a piston rod (Figure 4.2).

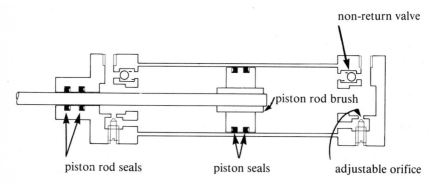

non-return valve

piston rod brush

piston rod seals   piston seals   adjustable orifice

Figure 4.1 *A double-acting cylinder or ram (pneumatic or hydraulic). The movement is cushioned at each end of the stroke, when the bush on the piston rod blocks the main fluid flow: for the last part of the travel fluid can pass only through the adjustable orifice. When the direction is reversed, fluid can bypass the orifice through the non-return valve.*

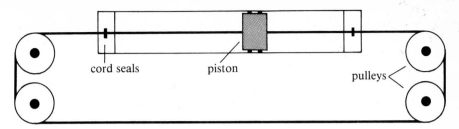

Figure 4.2 *A rodless cylinder.*

Among cylinders with piston rods, the two most common types are those of Figure 4.3(a) and 4.3(b): double acting, in which compressed air can be admitted to either end, and single acting, which exerts pneumatic force in only one direction, with a spring acting in the other. The latter has the advantage of needing a simpler valve and a single pipe, but is of restricted stroke and limited force in the inward direction.

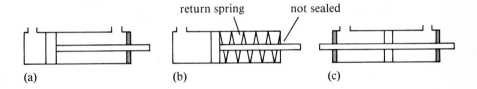

(a)   (b)   (c)

Figure 4.3 *Common types of cylinder (pneumatic or hydraulic): (a) double acting, (b) single acting, (c) equal area or double rod cylinder.*

At this point we may note some properties of all cylinders, hydraulic as well as pneumatic. The first is that they cannot resist torque about the long axis and so, if axial rotation is undesirable, an extra sliding guide (which may be a second, parallel, cylinder) must be used. (An exception is the first type of rodless cylinder.) Secondly, their ability to withstand side loads varies among different designs and, for a given ram, is least when the ram is fully extended. It is usually rather poor, and so designs like the first Unimates in which the piston rod supports a heavy sideways load are rare. Finally, when extended they are very long and therefore it is difficult to make a compact machine.

Figure 4.1 shows the main elements of a double-acting cylinder. A pneumatic actuator can also be made using a bellows; these are only suitable for short fat actuators of high force but limited stroke.

The other class of pneumatic actuator is the rotary actuator. A vane in a chamber rotates through an arc (Figure 4.4). The vane seal generates high friction and the angle of rotation is restricted. An alternative is a ram or pair of rams with a built-in rack and pinion.

PNEUMATIC VALVES

Pneumatic systems in general contain many kinds of valve, for limiting pressure to a safe level or keeping it constant or regulating flow, and other components such as filters and oil-mist lubricators. For the present purpose it is enough to consider just those used directly with actuators. Two functions must be provided: air must be switched to the correct side of the actuator; and the speed of travel must be controlled or at least limited, which is done by an orifice, usually on the exhaust side. The switching of air is done by solenoid-operated valves. The simplest valve, in which

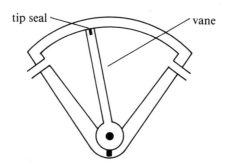

Figure 4.4 *Rotary actuator*

a single sealing ball, cone or plate is lifted off its seat by the solenoid, switches the flow through a single pipe on or off.

A form of construction which allows more possibilities is the spool valve, which is also common in hydraulics. An on/off two-port valve can be made in this way, but more interesting are valves with three or more ports, as they can perform the function of a group of on/off valves. A three-port, two-position spool valve acts as a changeover switch and can be used to control a single-acting cylinder in such a way that it is normally connected

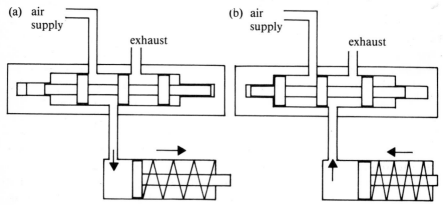

Figure 4.5 *Use of a three-way, two-position spool valve to control a single-acting cylinder: (a) with the spool to the right the cylinder is pressurized, (b) with the spool to the left the cylinder can exhaust (there is no position in which the cylinder is simply isolated, so there is no stable position except the ends of the travel).*

to the atmosphere but is switched to the air supply when the solenoid is active (Figure 4.5). One such valve replaces two on/off valves. To control a double-acting ram, two three-port or four on/off valves would be needed; instead, a single more complex valve can be used, and this is the usual method. This valve has three spool positions and five ports (sometimes the two exhaust ports are connected internally and brought out to a single external port, so the valve appears to have only four ports). It is shown in Figure 4.6.

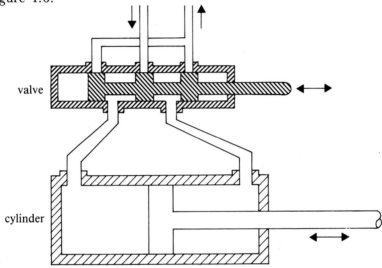

Figure 4.6 *Control of a double-acting cylinder by a three-position, five-port spool valve. With the spool in the central position (as shown) the cylinder is isolated; with it to the right pressure is connected to the left of the piston, and the exhaust or return line is connected to the right, and vice versa.*

Hydraulic actuation

Hydraulic power is used for the largest telemanipulators, some of which can carry a payload of several tons, because it is possible to generate an extremely high force in a small volume, with good rigidity and servo control of position and velocity. The high force is a consequence of the pressure at which hydraulic systems are operated, 130 bar (2000 psi) being common and higher pressures not unusual. Hydraulic actuation is also common for large industrial robots, although recent advances in electric motor design and gearing have made electric actuation competitive with or better than hydraulics in many cases, and most new industrial robots are electric.

The basic components such as cylinders and valves are similar to those used in pneumatics, so this section concentrates on the differences.

Hydraulic cylinders are usually double acting, although special forms exist, such as telescopic ones to give extra extension as in the body-tilting rams of tipper trucks. The equal area or double-rod cylinder of Figure 4.3 (c) has equal oil flows on the two sides (important in hydrostatic circuits, as described later) and generates equal forces in each direction.

Rotary hydraulic actuators are sometimes used in robotics. They are usually hydraulic motors, which work like a pump in reverse; several of the common pump mechanisms such as gears, pistons and rotating vanes are in use. Hydraulic pumps are described later.

HYDRAULIC VALVES

As with pneumatics a variety of valves exist for flow, direction and pressure control, and, as before, double-acting cylinders are controlled by four- or five-port spool valves. Hydraulic spool valves, however, are often used for speed control, so as well as simple three-position solenoid valves there are also servo valves. In a servo valve the spool position can be held at an accurately controlled postion at any point in its travel, and so flow as well as direction can be controlled by an electrical signal. A typical method of operation (Figure 4.7) is for a torque motor, whose movement over a few degrees is proportional to the input current, to regulate the differential oil flow through a pair of nozzles; this flow, much smaller than the main flow, determines the pressure on each end of the spool, and hence the spool movement. In effect the valve uses a hydraulic amplifier to boost the torque motor force to a level sufficient to move the spool. The ports and the profile of the spool are designed to make the relationship between the main flow and the spool position as linear as possible.

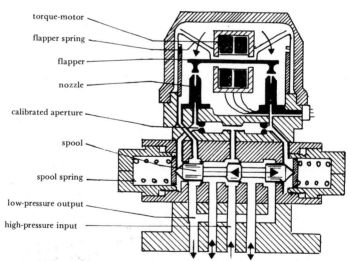

torque-motor
flapper spring
flapper
nozzle
calibrated aperture
spool
spool spring
low-pressure output
high-pressure input

Figure 4.7 *A cross-section through a hydraulic servovalve.*

Hydraulic servo valves are expensive. A simpler alternative, in which flow is electrically controlled but with less precision, is the proportional valve. Its construction is similar to a three-position directional spool valve, but the solenoids are designed so that when used with the appropriate amplifier the spool displacement is approximately proportional to an input voltage.

HYDROSTATIC CIRCUITS

Although servo valve velocity control is usual in industrial robots, it tends to be energetically inefficient. Whenever oil flows through a restriction under pressure, power is dissipated as heat. As well as wasting power, the oil heats up rapidly unless efficiently cooled. An alternative is to use a hydrostatic circuit in which fixed volumes of oil circulate in closed loops without passing through servo values. An example is shown in Figure 4.8, in which an equal area cylinder is driven by a pump. Unlike a conventional system, the hydrostatic method needs a pump for each actuator. Also, in practice the circuit is more complicated than that shown here, owing to the need for filtration, topping up and so on. Hydrostatic circuits have been advocated for walking robots, where efficiency is important.

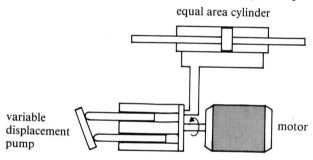

equal area cylinder

variable displacement pump

motor

Figure 4.8 *A hydrostatic circuit.*

HYDRAULIC PUMPS AND ASSOCIATED EQUIPMENT

The smallest pumps use a pair of closely meshing gears as shown in Figure 4.9; the oil is trapped in the spaces between the teeth and the casing and forced round the outside of each gear. These pumps are compact, cheap and simple.

At the highest pressures piston pumps are used. There are many designs. One, the swash plate pump, is shown in Figure 4.10. The pistons are arranged like bullets in the chambers of a revolver. As the cylinder block spins they are forced in and out by their contact with the inclined swash plate; they are connected to the inlet and outlet ports in turn by holes in the port plate. The swash plate pump, unlike the gear pump, is one of the pumps whose delivery rate can be varied by a control signal, in this case altering the angle of the swash plate, while the cylinder block rotates at constant speed. In this case it is called a variable displacement pump. There is usually an internal hydraulic feedback loop so that the swash plate angle can be changed easily despite the large forces on it. Many other types of pump exist. Figure 4.11 shows a variable displacement pump using radial pistons.

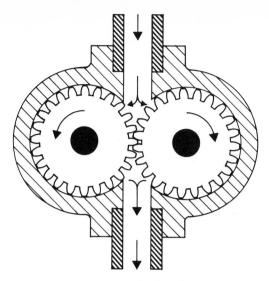

Figure 4.9 *A cross-section through a gear pump.*

A hydraulic pump is usually found as part of a pressure generation system or 'hydraulic power pack'. The basic features of such a system are shown in Figure 4.12. In a static installation the pump is usually driven by a mains electric motor, to which it is directly coupled. The motor and pump are bolted to the reservoir, a tank containing a coarse filter on the suction line and a screen to prevent any bubbles from the return line being sucked into the inlet. The relief valve limits the pressure to a safe value, returning any excess flow to the tank. Fine filtration is usually done on the return line; the filter is then subjected only to a low pressure.

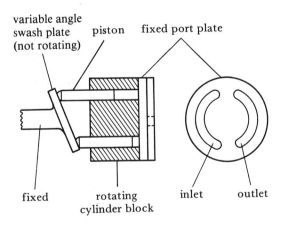

Figure 4.10 *A cross-section through a variable displacement swash plate pump.*

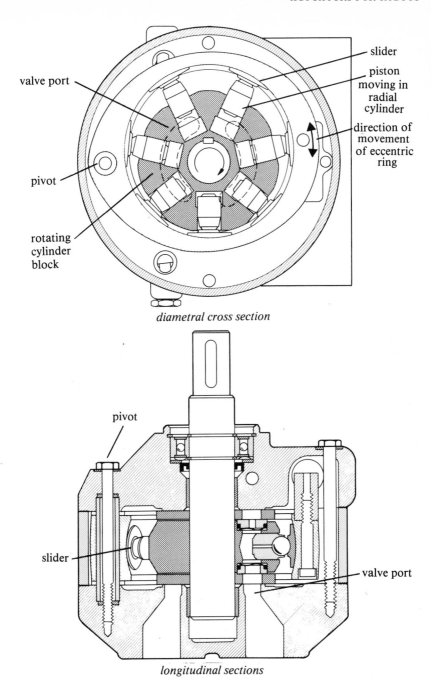

diametral cross section

longitudinal sections

Figure 4.11 *A radial piston variable displacement pump (courtesy of Danfoss Ltd) — the displacement is determined by the position of the eccentric ring powered by a hydraulic actuator; this can be servo-controlled in many ways, such as keeping the pump at constant pressure regardless of flow.*

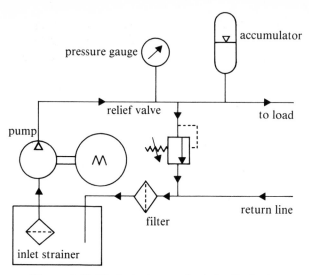

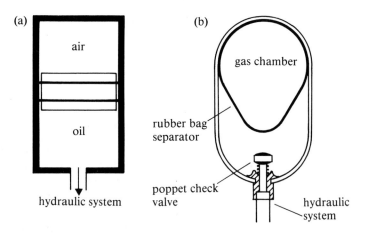

Figure 4.12 *The basic elements of a hydraulic circuit.*

An *accumulator* is an energy storage device using a pneumatic spring (Figure 4.13). It allows a load to draw for a short time, perhaps a fraction of a second or a few seconds, a peak flow greater than can be supplied by the pump.

Figure 4.13 *Two designs of a hydraulic accumulator.*

The working fluid is nearly always hydraulic oil, which is very like car engine oil since it has good lubrication properties and does not freeze easily or cause rusting. However, water and other fluids are not unknown, when flammability or leaks of a messy liquid must be avoided.

Electric actuation

Electric actuation is the dominant method for industrial robots. Most electric robots use servomotors; these can be divided into *DC servomotors,* which power the majority of existing robots, and a second class sometimes called the *brushless DC servomotor* and sometimes the *AC servomotor,* which is becoming increasingly popular. Electric actuation can also be done by *stepper motors,* for a limited class of robots, as discussed later.

DIRECT CURRENT SERVOMOTORS

These have been the most common kind of motor used in industrial robots until now. A rotor wound with an armature rotates in the magnetic field of a set of stationary permanent magnets. A commutator is needed to switch the current to the windings as the rotor spins. Some motor geometries are shown in Figure 4.14. Configurations intermediate between the disc and the bell, with a dished rotor, also exist. In a disc motor the torque-generating parts of the windings run radially; they are made as a printed circuit on a disc of fibre-reinforced epoxy resin.

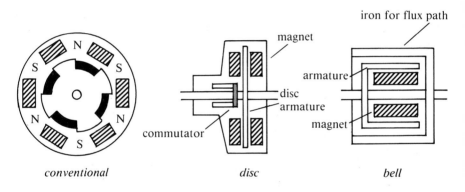

conventional   *disc*   *bell*

Figure 4.14 *Types of DC servomotor.*

The permanent magnets are generally made of either alnico (an aluminium-nickel-cobalt-iron alloy) or, when the resulting decrease in size justifies the cost, a mixture including cobalt and compounds of rare earth elements, particularly samarium. Magnets of ferrite (magnetic oxides of iron and other elements) also exist.

Disc and bell motors have low inertia and therefore high acceleration. A flat disc is less rigid than a dish or bell but easier to print with conductors. Conventional motors are the most able to withstand high torques and other stresses. The choice of motor may also depend on the space available, as disc motors are short and fat compared with conventional motors.

Commutation is done in the usual way, with graphite brushes on a segmented disc or drum.

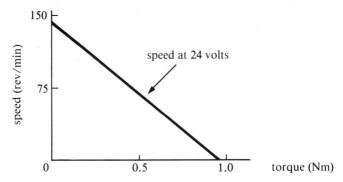

Figure 4.15 *The speed/torque characteristic of a DC servomotor.*

A DC servomotor is essentially a transducer of current to torque, so one of its important specifications is the torque constant k_φ, which is the torque per unit current and is measured in newton-metres per ampere. Another important characteristic is the graph of speed against torque at a fixed voltage; it is a straight line of negative slope (Figure 4.15). The full torque often implies a power consumption which would cause overheating if maintained for long, but which can be tolerated for short times. The equation of this line is

$$\omega = \frac{k_\phi - RT}{k_\phi^2} \tag{4.1}$$

where ω is the angular velocity, R the winding resistance and T the torque. The relationship between voltage and torque is

$$V = \frac{RT}{k_\phi} + k_\phi\omega \tag{4.2}$$

or, in terms of current I instead of torque,

$$V = RI + k_\phi\omega \tag{4.3}$$

These are steady-state equations. If the current is changing, a more exact form of equation (4.3) is

$$V = RI + L_\mathrm{L}\frac{\mathrm{d}I}{\mathrm{d}t} + k_\phi\omega \tag{4.4}$$

In all these equations k_φ is equal to E, the back electromotive force (e.m.f.). In an ideal motor we would have

$$V = E \tag{4.5}$$

It would also be possible to write a power equation

$$VI = T\omega \tag{4.6}$$

to express the fact that all the electrical power was converted into mechanical power.

The reason for introducing these equations is that the literature on servo systems treats servomotors using a bewildering variety of mathematical models, for example describing the back e.m.f. as a velocity feedback loop. The equations presented here should help to interpret such descriptions.

ALTERNATING CURRENT SERVOMOTORS (BRUSHLESS DIRECT CURRENT SERVOMOTORS)

If a permanent magnet DC servomotor is turned inside out so that the magnets form the rotor while the windings are fixed, then no commutator connection to the rotor is needed. Instead the current in the windings must be switched by electronic commutation, or must alternate, in synchronism with the motor rotation. The rotor may carry a slotted disc or a ring of magnets to operate optical or Hall-effect switches to synchronize the commutation.

AC servomotors have several advantages over DC commutator motors:

1) higher reliability;
2) low friction;
3) less electrical noise owing to elimination of the commutator;
4) no sparks;
5) longer life;
6) high torque to inertia ratio;
7) high torque throughout the speed range.

They require an electronic controller to generate the alternating coil currents. Figure 4.16 shows a range of AC servomotors.

Figure 4.16 *Indramat AC servomotors (courtesy of G.L. Rexroth Ltd).*

STEPPER MOTORS

The attractive feature of stepper (or stepping) motors is that as long as their rated torque is not exceeded their angular position remains known and so no joint angle sensors are necessary. Also, since the current to each coil is either on or off, simple switching circuits can be used to drive them.

Steppers are usually small and of limited torque and speed and tend to be confined to low performance robots for educational applications. Powerful steppers do exist and are used in machine tools and for the wheels of some mobile robots, but their power to weight ratio is too poor for manipulators. Hydraulic amplifiers are available in which an output shaft driven by a powerful hydraulic servo system follows the steps of an input shaft attached to a stepper motor, but this method is unsuitable for robot arm joints since the stepper's limitations of speed and acceleration still apply.

Figure 4.17 shows the principle of a stepper motor. The rotor is a set of permanent magnets or iron cores in the variable reluctance motor. A two-pole permanent magnet rotor is shown with a six pole, three-phase stator (a real motor would have many more poles). When any coil is energized the rotor has one stable position. In Figure 4.17(a) coil A is energized. When it is turned off and coil B is turned on, the stable position changes and the rotor rotates to align with it. There may be several sets of poles associated with phase A, alternating round the circumference with the poles of other phases. In this way a small step angle can be achieved; step sizes of a few degrees are usual. There are often four phases.

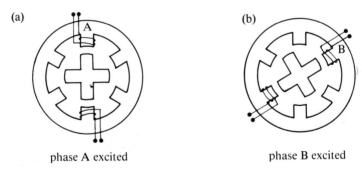

(a) phase A excited (b) phase B excited

Figure 4.17 *The principle of permanent magnet stepper motor operation.*

A stepper motor can be driven in discrete steps so that it comes to rest between successive steps, or the coil currents can be switched at an increasing rate so that the motor accelerates, the speed increasing by a definite amount at each step until some desired speed is reached, when the coil switching rate is kept constant. The stepper then acts as a kind of synchronous motor. Figure 4.18 shows how position and velocity change in the discrete step and continuous rotation cases. The jerkiness of the motion is governed by the ratio of the motor torque to the inertia of the load, and for large loads the motion may be quite smooth. Throughout this process of acceleration and high speed running, the step count can be maintained and the exact angular position and speed known. It is important not to step at a rate which would imply a load acceleration or deceleration, and therefore a torque, greater than the motor can supply.

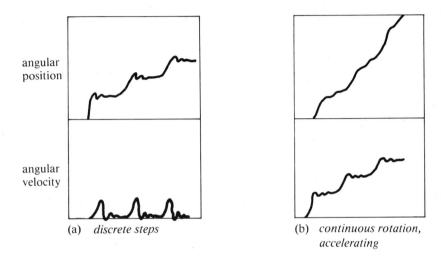

angular position

angular velocity

(a) *discrete steps*

(b) *continuous rotation, accelerating*

Figure 4.18 *Two modes of operation of a stepper motor: (a) discrete steps, (b) continuous rotation, accelerating.*

OTHER ELECTROMECHANICAL ACTUATORS

Electric motors are, of course, electromagnetic devices. However, other transducers of electrical to mechanical energy exist, and some are used as actuators in non-robotic applications; an example is the electrostatic loudspeaker. A list of phenomena usable in actuators is given below:

1) electrostatic force: electric field → force or displacement
2) piezoelectricity: deformation of a solid in an electric field
3) magnetostriction: deformation of a solid in a magnetic field
4) thermal expansion: bending of a bimetallic strip on heating
5) shape memory effect (SME): deformation of a solid on heating

Others are possible, particularly if chemical reactions and fluid states of matter are considered, but the ones listed are useful because they can produce mechanical movement in a solid structure very simply. The last two rely on electric heating; it is the change in temperature which produces the movement.

These phenomena have characteristics such as small movement, low force or long time constant which make them useless for powering the main joints of an industrial robot, but there are potential applications in robotics where they can be useful. These are where a very small mechanism is needed or very precise movements must be made.

Experimental robots have been made with both piezoelectric and shape memory actuators. The piezoelectric effect is so small that to get an appreciable movement two crystals, cut so that one expands while the other contracts, are glued together, like a bimetallic strip, so that the whole beam bends (Figure 4.19). This is called a bimorph. If three such beams are attached to each other at right angles, an approximately Cartesian three-dimensional micromanipulator can be made (Figure 4.19(b)).

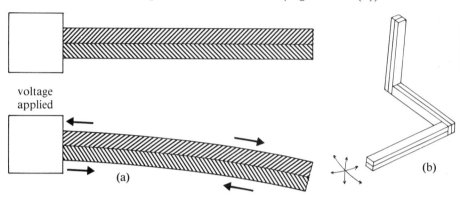

Figure 4.19 *Piezoelectric bimorph beam actuator: (a) principle of bimorph, (b) a three-axis actuator.*

The SME can generate larger and more powerful movements. Certain alloys called shape memory alloys (SMAs) show a phase transition between two crystal structures when mechanically deformed which can be reversed by heating above a certain temperature. Alloys of titanium–nickel (Ti–Ni) and copper–zinc–aluminium show the effect. If a piece of the Ti–Ni alloy is given its shape, heat treated at 750°K (about 500°C) and then allowed to cool, it can be deformed when cold and will retain its deformed shape, but on reheating it will spring back to the original shape; when deformed cold it retains the 'memory' of its original shape. This cycle can be repeated indefinitely.

The simplest SME actuator is a wire which has been made above the transition temperature (to speak of a single transition temperature is actually a simplification), allowed to cool and then plastically stretched. A piece of this wire can be used as a tension-generating rectilinear actuator (like a muscle). A current is passed through it so that resistance heating brings it up to the transition temperature, when it contracts to its original, pre-stretching, length; on cooling it expands again.

The resistivity of Ti–Ni SMA is only 50 times that of copper, so a high current density is needed to produce the heating. This leads to designs with long thin wires, which must then be bundled up somehow, by forming into a zigzag or helix. An example is shown in Figure 4.20. This is a bidirectional actuator: the right half pulls the 'piston' to the right and vice versa. The series connection of the elements increases the total resistance of the unit, matching it better to common voltage and current levels. An experimental actuator of this type produces a force of 4 N and a displacement of a few millimetres; the resistance of the whole arrray is about 10 Ω.

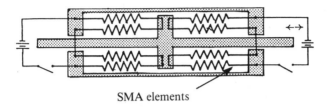

SMA elements

Figure 4.20 *A shape memory actuator.*

The speed of operation is limited by the cooling time, which depends on how efficiently the wire can lose heat. High speeds need thin wires; cooling can be improved by immersion in a medium such as water.

SME actuators are light and simple and can generate a high force. They are slow, however, and of low energetic efficiency, and so will be restricted to specialized applications.

Mechanical transmission methods

An actuator occasionally drives a joint directly. This has the advantage of introducing no backlash or elasticity and of minimizing complexity; if actuators were always small, light and powerful enough, direct driving would be almost universally used. However, actuators are usually heavier and bulkier than desirable, and of too low a force or torque, so mechanical transmissions are usual. The reasons for using a transmission are as follows:

1) to reduce the mass, volume and moment of inertia of the machinery at the end of the arm;
2) to match the speed and torque, or range of movement, of an actuator to its joint (reduction gearing and amplifying linkages);
3) to use a rotary actuator to power a prismatic joint or vice versa;
4) to perform mathematical or control functions — the two main examples are differential gears and parallelogram linkages.

This section describes some mechanisms commonly used for these purposes.

INERTIA REDUCTION

When an actuator is to be remote from the joint it drives, its power can be transmitted to the joint by one or more of these methods:

1) rigid links,
2) shafts,
3) chains, wires or tapes,
4) gear trains.

Some examples are now given. Figure 4.21 shows a common industrial robot construction in which a parallelogram linkage is used so that the actuator for axis 3 can be mounted as near the base of the robot as possible. Figure 4.22 shows the use of a pair of chains (Yaskawa Motoman) or a pair of linkages (ASEA IRb-6) to connect axis 4, the wrist pitch joint, to a motor mounted low down on the base. It is even possible to drive wrist roll and yaw from base-mounted motors; this is not generally done in industrial robots (although it is in telemanipulators) as it is difficult to make such complex transmissions with enough rigidity for an industrial robot. Various compromises are usual: for example, the wrist motors are often mounted near the elbow, driving the wrist through shafts and gears. The three wrist motors can be seen in a row on the back of the elbow of the KUKA IR 160/15 shown in Figure 4.23.

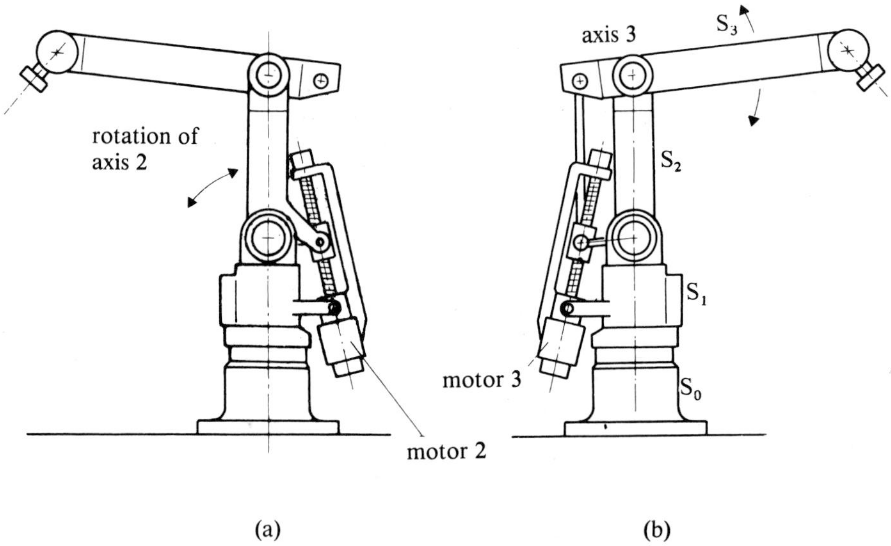

Figure 4.21 *Linkages for driving robot axes from base-mounted motors: (a) left side – drive to axis 2, (b) right side – drive to axis 3.*

SPEED REDUCTION

Electric motors rarely develop enough torque to drive a robot axis directly. The most common approaches to gearing are (a) to build a coaxial reduction gear onto the motor and (b) to use a screw and nut.

Reduction gears use epicyclic (planetary) gear stages (Figure 4.24) or

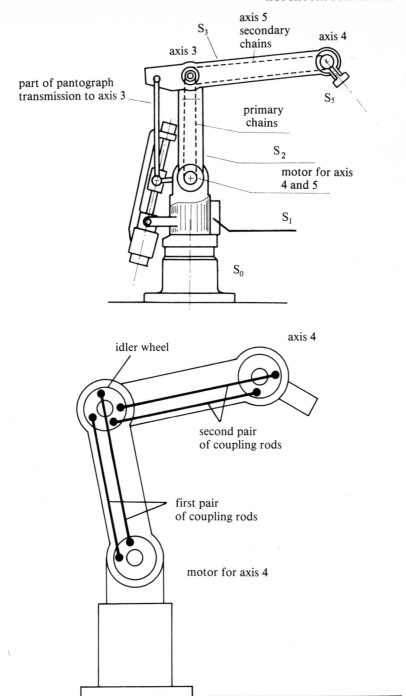

Figure 4.22 *Chains and linkages for remote driving of robot joints.*

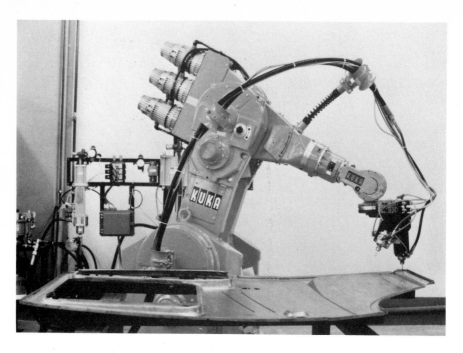

Figure 4.23 *A robot (the KUKA IR 160/15) with indirect drives to its three-axis wrist. The wrist motors can be seen at the back of the 'elbow' (courtesy of KUKA Welding Systems & Robots Ltd).*

Figure 4.24 *An epicyclic reduction gearbox. Left — fixed outer ring gear. Centre — cage or 'spider' which is attached to the output shaft and carries three planet gears; the central gear driven by the input shaft can also be seen. Right — the assembled reduction gearbox; this is actually one stage or layer of a gearbox made up of a stack of such stages.*

harmonic drives and their relatives. Figure 4.25 shows one of these, the cycloidal speed reducer. The principle is that, if a disc rolls inside a slightly larger ring, the speed with which it rotates about its own axis is much lower than that with which its centre orbits the central input shaft. Therefore if an output shaft is connected to the disc so as to pick up its rotation (as opposed to its orbital motion) the output speed will be a small fraction of the input speed. This connection is made by a ring of pins which penetrate holes in the disc (Figure 4.25(b)). The disc must roll on the outer ring without slipping; this is ensured by gear-tooth-like lobes of cycloidal profile which mesh with rollers fixed to the ring.

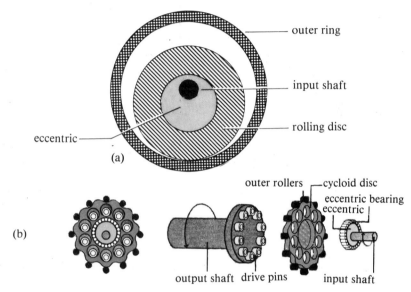

Figure 4.25 *Cycloidal speed reducer: (a) the principle of the reducer, and (b) a schematic drawing.*

A related device is the harmonic drive (Figure 4.26). Again the input shaft carries an eccentric or, rather, a cam, this time elliptical. The casing has gear teeth corresponding to the rollers of the cycloidal drive. The rolling disc is replaced by a flexible toothed membrane which takes up the shape of the elliptical cam and to which the output shaft is attached. It has slightly fewer teeth than the casing. As before it rolls round the outer ring, rotating much more slowly than the input eccentric.

The screw and nut used for speed reduction uses a recirculating ball nut for low friction. The output is rectilinear motion, and so if a revolute joint is to be driven a rack and pinion or a crank is needed to convert the motion back into a rotation. This is the arrangement of Figure 4.21. The main advantage of this rather roundabout approach is that it reduces the torque in the shaft of the joint. It also allows more choice of where to put the motor; otherwise several coaxial motors would be needed in some designs. It may also be easier to reduce backlash and to increase stiffness, since coaxial reduction gearboxes tend to have a relatively large amount of play in the gears.

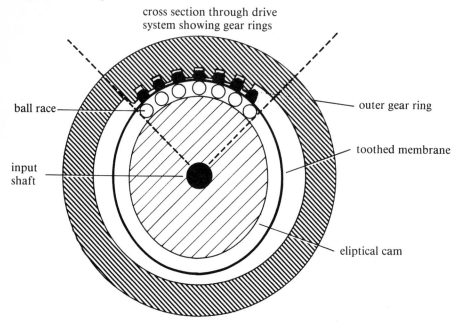

Figure 4.26 *Harmonic drive.*

ROTARY/RECTILINEAR MOTION CONVERSION

Apart from the screw and nut (ball screw, leadscrew) the main methods are the rack and pinion, the chain (or wire or tape) and pulley, and the crank. All these methods can be used for rectilinear to rotary as well as rotary to rectilinear conversion, although the screw and nut is not normally used for rectilinear to rotary conversion because friction makes it inefficient or even impossible to drive in this direction.

DIFFERENTIALS AND PARALLELOGRAM LINKAGES

The use of differentials to compensate for unwanted movements was discussed in the section on wrists in Chapter 2.

Linkages of rods not only can allow the actuator to be remote but can be used to preserve orientation over several joints; this was mentioned in Chapter 2 and is true of the transmissions of Figures 4.21 and 4.22, in which the orientation of segment S_3 and the wrist are not affected by motion of the intermediate joints.

Bibliographic notes

A useful introduction to pneumatics and hydraulics is McCloy and Martin (1980). Stepping motors are covered in some detail in Lhote *et al.* (1984), which explains the difference between permanent magnet and variable reluctance motors; it also discusses DC servomotors and mechanical transmissions.

Shape memory actuators are described in Hirose *et al.* (1985).

Chapter 5

Sensing for Robots

The terms 'sensor' and 'sensing' refer to the detection or measurement by a robot's controller of any physical state or quantity, from the state of a switch to a television image.

In general, a sensor is a system in its own right, although it is a trivially simple one such as a microswitch in some cases. This system consists of a transducer, or a combination of transducers, together with some electronics to produce a signal suitable for interfacing to the robot controller. A transducer in this context is a device which produces an electrical signal which is a measure of a physical quantity; an example is a potentiometer which produces a voltage proportional to a shaft angle. The output of a transducer is usually either a voltage or a digital signal. A sensor such as a rangefinder might use several transducers and might include some computational capacity so as to give a range measurement, in the form of a digital number, as its output. It is often unimportant to distinguish between transducers and sensors, especially since transducers increasingly have some signal processing electronics built into them.

Transducers are never perfectly accurate, and their specifications include several quantities expressing aspects of this. Figure 5.1(a) shows the output of an angle transducer and defines 'offset'. *Non-linearity* is the departure of the actual voltage–angle characteristic from a straight line. There is also an error in the voltage–angle proportionality constant, since the average slope of the measured characteristic differs from that of the ideal line. A possible error which is not shown is *non-repeatability* due to, for example, backlash in gearing.

Apart from errors, a most important parameter is the *resolution*, which is the smallest interval which can be distinguished; it is often expressed as the *fractional resolution*. The resolution of the sensor in Figure 5.1(a) is infinitely good (in practice limited by electrical noise); Figure 5.1 (b) shows a sensor with a finite resolution, as is always the case with digital instruments.

The rest of this chapter describes the sensors most commonly used in robotics.

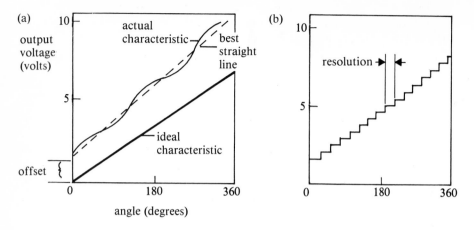

Figure 5.1 *Characteristic of a transducer (shaft angle to voltage): (a) infinite resolution, (b) finite resolution.*

Joint angle

The fractional resolution needed varies from about 1 in 200 for a small, low accuracy robot to perhaps 1 in 10^5 for a large, high precision assembly robot. It is almost impossible to measure an angle of 1 in 10^5 directly with a compact and robust sensor, so if such a resolution is needed a transducer of poorer resolution is fitted somewhere in the gear train driving the joint; it will make several revolutions for every one of the joint, so a revolution count must be kept. Such an arrangement may be regarded as a combination of a fine sensor (the transducer itself) and a coarse one (the revolution count).

Since digital methods are universally used for robot controllers, and the data bus width of most small computers is 8, 16 or occasionally 12 bits, it is natural to try to work with sensors giving 8-, 12- or 16-bit outputs. These correspond to the resolutions given in the table below which also shows the positional resolution achieved at the end of a link 10 cm or 1 m long attached to the joint, for an angular range of 270° (about the limit for many joints).

Number of bits	Fractional resolution	Angular resolution (degrees, for 270° range)	Positional resolution at 10 cm (mm)	Positional resolution at 1 m (mm)
8	1 in 256	1.05	1.8	18
12	1 in 4096	0.066	0.11	1.1
16	1 in 65536	0.0041	0.0072	0.072

Before giving a list of angle sensing methods, it may be noted that if the joint is driven by a stepper motor then, provided that a count of the steps is kept, no angle sensing is needed, although if there is any danger of the motor missing a step (as may happen under excessive load) a separate angle sensor is advisable. Some small cheap arms rely on step counting.

POTENTIOMETERS

Potentiometers are relatively cheap and simple and can have almost infinite resolution; their accuracy is limited more by non-linearity and noise. Linearities of perhaps ±0.05% (1 in 2000) can be achieved by hybrid track potentiometers in which a wound wire element determines the linearity and a conductive plastic coating on the winding bridges the gap between individual turns and so avoids the staircase effect seen in the output of ordinary wirewound potentiometers.

Potentiometers are rarely used in robots. Apart from their linearity not being good enough for the highest accuracies, they need more individual calibration than other transducers. An added weakness is that the output voltage is proportional to the supply voltage, which must therefore be extremely well regulated.

RESOLVERS AND SYNCHROS

Resolvers and synchros are transformers having two or three fixed windings and a rotor winding; Figure 5.2 shows the basic form. The rotor coil carries an alternating current at a few kilohertz; the amplitude of the alternating voltage induced in the static coils depends on the rotor angle. For a resolver the amplitude of the voltages in the two coils is proportional to $\sin \theta$ and $\cos \theta$, respectively, where θ is the shaft angle. The synchro is a three-phase device, designed originally to drive a similar three-phase indicator. Resolvers are used in some robots, e.g. the ASEA IRB6. An angular resolution of a few arc minutes (one in a few thousand) can be achieved. The signal is harder to convert into a digital measure of angle than that of some transducers, but resolvers are robust and the signal is resistant to interference.

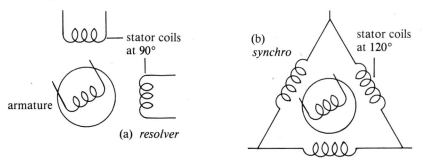

Figure 5.2 *Principle of operation of (a) resolvers and (b) synchros.*

As with potentiometers, linearity is more of a limit on accuracy than resolution.

INCREMENTAL AND ABSOLUTE ENCODERS (USUALLY OPTICAL)

Incremental and absolute encoders are truly digital devices. They can be made with extremely high resolution and linearity and do not depend on a precisely regulated supply voltage.

Figure 5.3 shows the principle of both types. In each case the light and dark sectors on a disc are detected photoelectrically. Variants exist in which the sectors are detected by transmission and by reflection; some use a drum instead of a disc, or use brushes to detect contact with conducting and insulating sectors.

two sensors
one-quarter step
out of phase

four bit
grey code disc

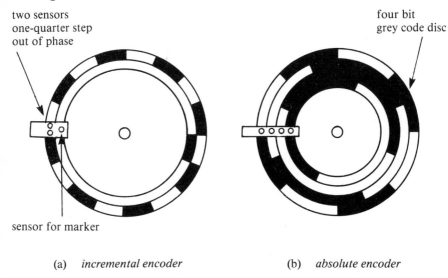

sensor for marker

(a) *incremental encoder*   (b) *absolute encoder*

Figure 5.3 *The principle of (a) incremental and (b) absolute encoders.*

With an incremental encoder the output, on each of two wires, is a series of pulses as the disc rotates. These pulses must be counted, starting from a known initial position, to give the total angle turned through. The two photocells are a quarter of a step out of phase; this enables the direction of rotation to be determined (Figure 5.4). A third photocell generates a marker pulse once per revolution. An incremental encoder is relatively simple and has only three output connections regardless of its precision but needs pulse counting, and there is a danger that the count might be lost in the event of a power failure or computer restart. Incremental encoders are commonly made with up to 5000 pulses per revolution.

The alternative is the absolute encoder which has the advantage of giving at all times a complete measurement of angle, in the form of a parallel binary signal, with no counting. Its disadvantage is that it needs a photocell

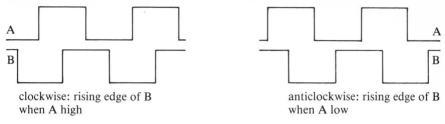

clockwise: rising edge of B
when A high

anticlockwise: rising edge of B
when A low

Figure 5.4 *Phase relationship of pulse trains from an incremental encoder in the two directions of rotation.*

and pattern ring on the disc for every bit in the word, so a 12- or 14-bit encoder is a complex device, and it also needs a cable with many conductors and large connectors. Absolute encoders are more expensive than incremental ones.

The disc of an absolute encoder normally uses Grey code in which only one bit changes at a time; this minimizes the chance of an error when the encoder is very near a transition between sectors.

For both incremental and absolute encoders, above about 12 bits the sectors become so narrow that detection is difficult unless the encoder is very large. Some encoders use analogue methods to measure angular movements smaller than the narrowest sectors on the disc.

Encoders are more delicate than resolvers and the output signals more liable to interference.

Joint angular velocity

An incremental encoder makes a good speed sensor since the pulse rate from it is proportional to angular velocity. Speed can in principle be measured by differentiating the output of any position transducer (usually by software rather than by a circuit), but in practice this tends to give poor accuracy. Dedicated sensors for angular speed are available, in general called tachometers. Apart from pulse counting the main method is to use a small DC or (occasionally) AC generator. Up to some limiting speed, the output voltage is proportional to angular velocity.

Rectilinear position*

Straight-line position can be sensed using a straightened-out version of any of the angular sensors just discussed. Rectilinear potentiometers are made with strokes up to about a metre. A rectilinear relative of the resolver or synchro is the linear variable differential transformer (LVDT) (Figure 5.5) in which the moving element is an iron core which couples a primary coil

---

* The term 'rectilinear', and not just 'linear', is used for motion in a straight line, since 'linear' is used for mathematical functions such as the output characteristic of a transducer.

to one or two secondary coils; the difference in the amplitude of the voltages induced in the two secondaries is proportional to the displacement of the core. A typical linearity is ±0.15% and strokes of up to 10 cm are common.

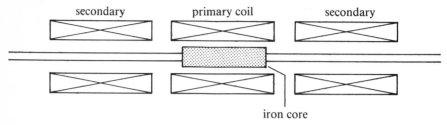

secondary   primary coil   secondary

iron core

Figure 5.5 *A linear variable differential transformer (LVDT).*

Rectilinear potentiometers and LVDTs are quite appropriate for measuring displacements of a few millimetres or centimetres and are suitable for use in wrists and grippers, but for a prismatic arm joint whose stroke may be a metre or more they are too bulky and delicate or cannot be made long enough while remaining accurate. Therefore rectilinear movements are often measured by angular sensors coupled by a rack and pinion or equivalent gearing.

Force and torque

In manipulation robots the main requirement for force or torque sensing is at the wrist, where it is used to sense forces during assembly or the force exerted on a workpiece by a tool such as a grinder held by the robot.

A second place for force measurement is in the jaws of a gripper. A third category is the sensing of forces, torques and stresses at various points in the arm to determine whether safe limits are being exceeded.

STRAIN GAUGES

Practically all force sensors measure the elastic deformation of a solid. In the simplest example a spring is compressed or stretched and its length change measured with a potentiometer or LVDT. This is suitable for small forces, as the spring can be made arbitrarily weak. It clearly involves a mechanical assembly and is therefore relatively complex. The preferred method for force sensing, provided that the forces are large enough, is to bond a strain gauge to a structural element which experiences compression or tension. This works because strain gauges can sense the extremely small dimensional changes in a structure when it is loaded. It has the advantage of being very simple, requiring no additional mechanism; however, the use of strain gauges needs care, as the leads are delicate and the signal is very small.

What a strain gauge actually measures is the change in length of the surface to which it is bonded; this can be an extension or a contraction. (*Strain* is defined as the fractional change in length.) By shaping the structure appropriately, the surface length change can be made a good measure of compressive or tensile force, or shear force, or bending force or torque, since, as can be seen from Figure 5.6, these all lead to local compressions and stretchings.

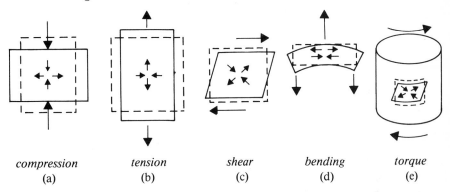

compression	tension	shear	bending	torque
(a)	(b)	(c)	(d)	(e)

Figure 5.6 *The local contraction and extension produced by a variety of types of force. The external arrows shown the applied force or torque; the small internal arrows show the local directions of movement (strain).*

Figure 5.7 shows some examples of structures designed to convert various forces and torques into contractions and expansions suitable for sensing by strain gauges. That of Figure 5.7(a) illustrates the fact that, even though a single gauge is in principle sufficient to sense a force in one direction, in practice a more complicated arrangement is often used, and so four gauges can be built into a bridge circuit. This increases sensitivity and reduces the effect of resistance changes due to extraneous sources such as temperature changes. More complex structures allow the simultaneous measurement of forces and torques in several directions.

Most strain gauges consist of a zigzag pattern of copper–nickel alloy etched on a polyester film (Figure 5.8). A typical resistance is 120 Ω, and a strain of 3% or 4% can be measured. The ratio of the fractional resistance change to the strain is called the gauge factor and is typically 2, in which case a 1% strain produces a 2% resistance increase. Gauges are made to match the thermal expansion coefficients of common metals such as mild steel and aluminium. Semiconductor strain gauges also exist and produce a larger signal but have poorer linearity.

PIEZOELECTRIC FORCE TRANSDUCERS

A compressive force can be measured by the piezoelectric voltage generated between the opposite faces of a crystal of certain substances such as quartz when it is compressed. As such crystals have little ability to support bending or tension they must be built into a suitable structure, such as a rod sliding inside a tube, which isolates the crystal from all forces except compression.

The piezoelectric effect also occurs in certain polymers which can be made in the form of a flexible membrane. These enable touch sensors of a variety of types to be made, as explained in the section on touch sensing.

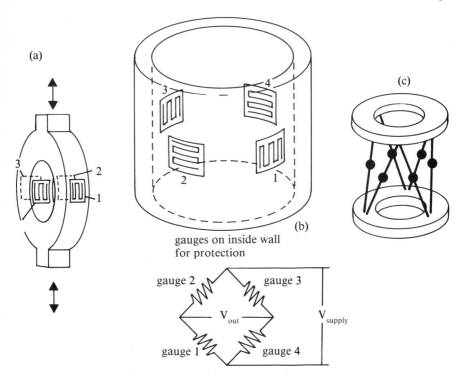

gauges on inside wall
for protection

Figure 5.7 *Strain gauge applications: (a) load cell for one direction of compression or tension; (b) another load cell and its bridge circuit; (c) six load cells — this measures three components of any force or torque applied between the two rings.*

Proximity sensing and range measurement

Proximity sensing is the detection of the presence of an object within a certain sensitive volume. It is used in collision avoidance, detecting whether an object is near a gripper, or between its jaws, and for safety barriers;

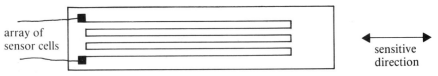

array of
sensor cells

sensitive
direction

Figure 5.8 *A metal foil strain gauge.*

other uses are possible. Proximity detection uses sensors similar to those for short-range distance measurement; the main ones are ultrasonic and optical, often infrared.

A very simple example of object presence detecion is the use of light beams between the jaws of a gripper.

Infrared optical proximity sensors can define a sensitive volume by the intersection of the field of view of the emitter and the receiver; the principle is shown in Figure 5.9. An object within the cross-hatched area will reflect light from the transmitter to the receiver. Very dark or shiny objects may not be detected. There may be several emitters or receivers to shape the sensitive volume, or to define more than one volume. An example of a system using this method is shown in Figure 5.10. It is a 'cuff' surrounding part of a manipulator to detect impending collisions. The main problem with optical sensors is interference by background lighting; this can be minimized by infrared filters and by detecting a pulsed signal from the emitter. Pulsing allows a much higher peak power to be emitted than if the emitter is powered continuously. The emitter is typically a gallium arsenside diode emitting at a wavelength of 940 nm; the pulse current can be several amperes for a few microseconds. The receiver can be a PIN photodiode housed in a black moulding which is transparent to infrared but absorbs visible light.

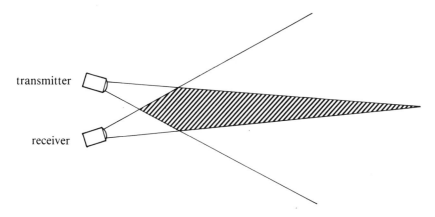

transmitter

receiver

Figure 5.9 *An optical proximity sensor.*

Ultrasonics is a good medium for object detection and range measurement up to a few metres; it is almost universal in mobile robots. Pulsed sonar at the commonly used frequency of 40 kHz has a distance resolution of about 1 cm, and a typical pulse repetition rate is 10 pulses per second. The transmitters and receivers, normally piezoelectric, are small and cheap and so can be used in large numbers. Figure 5.11 shows the type of transducer commonly used. Other kinds of proximity sensing, e.g. with capacitance, are sometimes used.

Figure 5.10 *An infrared proximity sensing 'cuff' surrounding a segment of a manipulator (courtesy of C.M. Witkowski).*

Figure 5.11 *Ultrasonic transducers; the transmitter and transducer are almost identical — these devices are tuned to a frequency of 40 kHz.*

Touch sensing

Several kinds of sensing fall under the heading of touch. First of all, there are several qualities which can be sensed, such as contact, force, displacement, shape and roughness. Then there is the question of whether the quality is sensed at a single point or at many points on a surface, typically arranged in a square grid. In the second case touch sensing becomes a kind of imaging, and the processing of the signals has much in common with vision.

Touch sensing is found in two main contexts: the touch between the jaw of a gripper and an object; and the contact during an assembly operation between the held object and the fixed assembly (the peg-in-the-hole problem). In the second of these it largely reduces to the sensing of wrist forces, which will not be discussed here.

In the case of object-to-gripper touch, the crudest kind of sensing is the detection of contact and can be done by a microswitch (Figure 5.12(a)). Another simple touch measurement is that of object size by jaw separation; this requires sensors of jaw position (Figure 5.12(b)). The force exerted can be measured by placing strain gauges in the linkage driving the jaws. Another kind of single-point touch sensing is the whisker or touch probe. Many forms are possible; in the one shown in Figure 5.12(c) the displacement of the probe tip in three dimensions is sensed. Such a probe can be used for seam tracking in some circumstances, or for following the edge of an object.

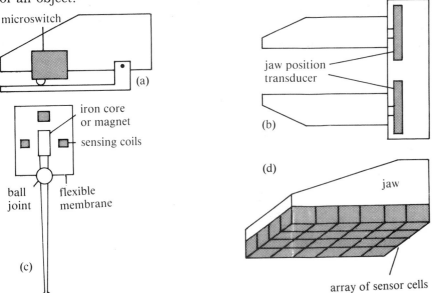

Figure 5.12 *Touch sensors: (a) object-contact sensor in jaw of gripper, (b) LVDTs or similar transducers for measuring jaw separation giving object size, (c) a 'whisker' probe, and (d) array of touch sensors for imaging.*

Figure 5.12(d) shows how touch sensors can be arranged in a square array so that some impression of the shape of an object can be obtained and a measure of its position within the jaws. Each element may be a simple contact sensor, or a force or displacement transducer. An array of the last-named can give a three-dimensional image of the part of the object touching the jaw, but this approach has had little success so far.

One of the problems with touch sensors is making them small, robust and reliable. Many methods have been tried; some are described below.

RESISTANCE-BASED TOUCH SENSORS

Elastic conductive materials, such as the carbon-loaded sponge in which integrated circuits are sometimes packed, register a resistance change when compressed. Unfortunately, this potential basis for touch sensing suffers from low sensitivity, non-linearity, noise, drift, a long time constant, hysteresis and poor fatigue resistance and so is of limited use.

Another resistance method uses bundles of carbon fibres; when compressed their resistance decreases.

INDUCTANCE AND CAPACITANCE

An array of LVDTs can give an accurate depth image of an object pressed against it, but the LVDTs are expensive and their bulk leads to poor spatial resolution since their centre to centre separation can be no less than a few millimetres. Inductive displacement transducers can also be made using printed circuit coils on an elastic membrane (Figure 5.13(a)). Similar methods can be used to make a rectangular array of capacitor plates (Figure 5.13(b)). Such sensors, whether capacitive or inductive, rely on the deformation of the elastic layer between the two sides. It must be quite thin to give good coupling, so the sensor is better adapted to detecting contact or force than to generating depth information. The signal from each pair of plates or coils is small. Another problem is that, since adjacent cells in the array are not mechanically isolated, a pressure at any point will produce signals in neighbouring cells.

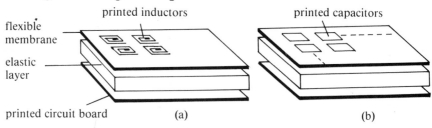

Figure 5.13 *Inductive and capacitive touch sensor construction. The three layers are shown separated for clarity. The printed circuit board has a pattern of coils or plates etched on it matching that on the flexible membrane.*

PIEZOELECTRIC TRANSDUCERS

Certain polymers such as polyvinylidine fluoride (PVDF or PVF_2) show the piezoelectric effect. They are also elastic and can be made in the form of a membrane; when the membrane is squeezed or stretched a voltage appears between one side and the other. This effect in PVDF is a few tenths of a picocoulomb per newton (the size of the effect is expressed as the charge generated per unit force applied), which is comparable with that in quartz.

In principle a touch sensor can be made in the same way as those of Figure 5.13. However, a fundamental problem is that the piezoelectric signal tends to be masked by a pyroelectric signal: materials such as PVDF produce a charge of a few nanocoulombs per square centimetre per degree if the temperature changes, as in general it will when the membrane touches an object.

Both the pyroelectric and the piezoelectric effects are used in human skin to good effect, so rather than simply eliminating the pyroelectric effect by thermal insulation a sensor can be devised to measure both effects (Figure 5.14). The outer layer is the thermal sensor; the inner layer, insulated thermally but not mechanically by the elastic layer, is the mechanical sensor.

As with a capacitance sensor array, adjacent cells are mechanically coupled, although this can be reduced, at the cost of added complexity and reduced strength, by cutting slots between the cells.

A further possible structure arises for piezoelectric elastomers because they can act as ultrasonic transducers, just as a quartz crystal can. This allows a design in which the depression of the outer layer of a touch sensor is measured by timing the flight of acoustic pulses from one side of an elastic block to the other (Figure 5.15, which shows just one cell of an array). This gives a direct measure of displacement, whereas the sensor of Figure 5.14 actually measures the force on the piezoelectric membrane below the elastic layer, which is only indirectly related to the displacement caused by the touched object.

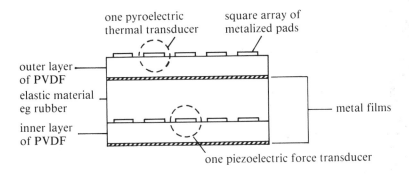

Figure 5.14 *The design of a piezoelectric touch sensor using polymer membranes.*

THERMAL TOUCH SENSING

The human thermal touch sense which enables us to distinguish between materials by how 'warm' or 'cold' they feel can be imitated by a sensor for robots. The sensor takes the form of an array of cells each consisting of a heat source, a layer of material such as silicone rubber, and a temperature transducer such as a thermistor. When any cell is in contact

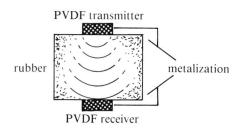

PVDF transmitter

rubber

metalization

PVDF receiver

Figure 5.15 *The design of a time-of-flight ultrasonic piezoelectric polymer touch sensor.*

with a material at room temperature, heat is conducted from the heat source (such as a power transistor) to the touched object through the rubber layer. The temperature in this layer near the contact surface is a measure of the thermal conductivity of the touched object. A rectangular array of cells can produce a touch image in the same way as other touch sensors.

Thermal touch sensors, currently in the form of small arrays such as 2 x 10 or 10 x 10, are under development at Wollongong University, Australia. One problem is that the thermal time constant of the sensors made so far is several seconds; it is hoped that this can be reduced by miniaturizing the transducers.

OPTICAL METHODS OF TOUCH SENSING

All the methods of sensing with an array of cells have the problem of needing a large number of wires to carry the signals from the cells, apart from mechanical problems. A different approach, although not entirely free of robustness problems, is to use optical imaging methods. An example is shown in Figure 5.16, in which contact by an object deforms a flexible membrane into contact with a thick window within which light is chan- nelled by total internal reflection. The window is viewed from the other side by a miniature television camera. In the absence of an object the field of view is dark, but contact by an object allows light to pass out of the window and to be diffusely reflected by the membrane, producing a light patch which is an image of the region of contact.

SLIP

If a held object starts to slip through the jaws of a gripper, this can be sensed by spring-loaded friction wheels driving angular sensors (Figure 5.17), but this is not often practicable. Another method is to build microphones into the jaws; if the held object is not too smooth, the vibrations produced as it slips over the face of the jaw can be sensed.

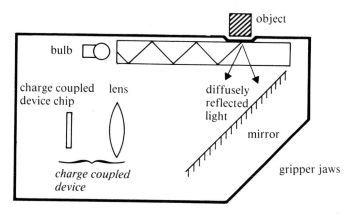

Figure 5.16 *Optical imaging touch sensor.*

The basic problem is that, although it is not too hard to detect slipping alone, if sensors of contact, force or object shape also have to be incorporated there is little room for extra sensors.

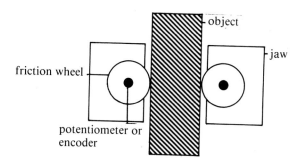

Figure 5.17 *A cross-sectional view of gripper jaws: a slip sensor.*

Vision

Robot vision is an enormous subject, particularly since it overlaps or makes use of fields such as pattern recognition and scene analysis which have many other applications. Consequently work on remote sensing, medical imaging, military target recognition and optical character recognition may be relevant to it. Also, advances in robot vision are intimately bound up with developments in computer hardware, and much computer development is driven by the need for faster image processing.

Some applications of vision in robotics are listed below:

1) detecting object presence or type,
2) determining object location and orientation before grasping,
3) feedback during grasping,
4) feedback for path control in welding and other continuous processes,
5) feedback for fitting a part during assembly,
6) reading identity codes,
7) object counting,
8) inspection, e.g. of printed circuit boards to detect incorrectly inserted components.

VISION HARDWARE

Vision systems are occasionally supplied by the robot manufacturer and integrated with the controller, but usually are separate, with an interface to the robot controller. Figure 5.18 shows the basic organization of a vision system. Not all the elements must be present, and some may be repeated.

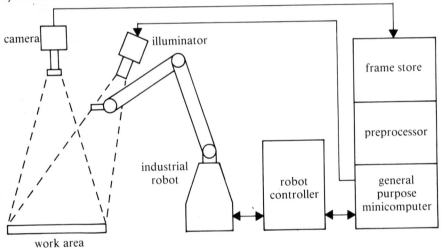

Figure 5.18 *The organization of a robot vision system.*

Television cameras

The simplest type is the linescan camera in which the sensitive device is a semiconductor chip containing an array of photodiodes or charge-coupled device (CCD) detectors in a straight line; there may be from 64 to several thousand detector elements. A 64-element photodiode array is shown in Figure 5.19. To produce a two-dimensional image a linescan camera relies on mechanical scanning for the second dimension. This can be provided by the movement of a continuous sheet or ribbon of material to be inspected, or of a conveyor carrying objects to be imaged (Figure 5.20).

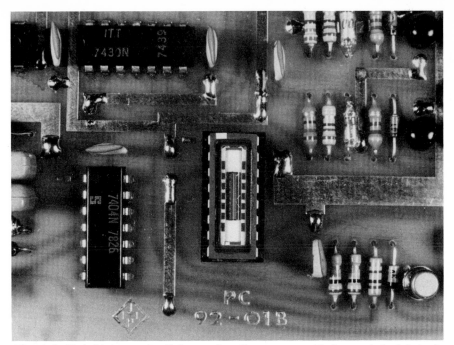

Figure 5.19 *A 64-element photodiode array for a linescan camera (the integrated circuit is in the centre of the photograph).*

Two-dimensional cameras use either a tube like that in a broadcast camera (a vidicon for normal conditions) or a semiconductor chip, usually a CCD, with a rectangular array of detector cells. For robotic purposes, the rectangle usually has a few hundred elements on each side: examples are 488 × 380 and 244 × 244.

If a vidicon camera is used it is usually a standard broadcast/closed-circuit television camera which produces 625 (in Britain) lines; the number

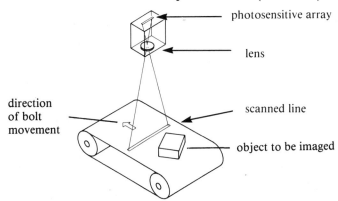

Figure 5.20 *Use of a linescan camera, with mechanical movement to generate the second dimension of a two-dimensional image.*

99

of picture elements ('pixels') in the line is determined by the digitizing interval and is chosen to be a convenient number, often 512. Sometimes some lines at the edge of the image are not digitized, or some blank lines are added, to make their number up or down to a convenient figure.

As in the linescan camera, a two-dimensional camera produces an analogue voltage signal which is a measure of the light level. Sometimes this is converted to a binary black or white signal by a comparator or discriminator circuit; otherwise it is converted to a digital signal of typically 6 bits (64 grey levels) or 8 bits (256 levels). The analogue signal is too noisy and limited in dynamic range to be worth digitizing with greater precision.

Colour has hardly ever been used in robotics, but has advantages or is essential in some applications: an example would be checking component colour codes.

Illuminator

The problem of discriminating an object from its background is made easier with proper illumination; e.g. the backlighting of opaque objects gives almost perfect contrast. However, it is often not practicable to provide special lighting.

Illumination may play a more active role in imaging: a special pattern such as a grid or stripe may be projected onto a scene using fan-shaped beams formed by slits or a mirror scanner. If the camera views the scene obliquely, depth can be estimated by triangulation from the position of the bright lines in the image. Figure 5.21 shows how a camera picks up the image of a light stripe projected across a joint between two plates. The image of the stripe contains a corner from whose position can be calculated the location in three dimensions of the joint.

Framestore

A framestore is a block of semiconductor memory which can be addressed in synchronism with the scanning of the camera and can hold a television frame. It has to be fast enough to accept a digitized television signal. For example, a 512 x 512 image received in 1/50 s implies that each memory location must be addressed and written to in $1/(50 \times 512 \times 512)$ s or about 70 ns. Once data are in the framestore they can be read, either as part of the memory of an ordinary computer or by special image processing hardware. The special hardware usually passes its results to a general purpose computer and in that case is called a preprocessor.

A framestore can often be written to by the computer as well as by the camera, so that a television picture with added graphics can be displayed on a monitor, to show how the results of various stages of computation compare with the original data.

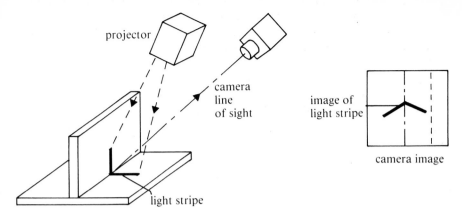

Figure 5.21 *Structured light: the camera detects a bar of light projected on the workpiece.*

Preprocessor

There is so much information in a television image that an ordinary minicomputer is too slow to do more than the simplest operations at rates useful for robotics, and so is often preceded by a preprocessor which uses dedicated circuits to do operations such as spatial filtering, edge detection, thresholding and calculating image statistics (such as the distribution of grey levels).

General purpose computer

A general purpose computer is a conventional minicomputer. If there is no preprocessor then it must do the low level operations such as thresholding and edge detection as well as the later stages; otherwise it can work with preprocessed, and therefore much simplified, data.

An alternative to the conventional preprocessor and minicomputer architecture is to use computers of radically different design, intended specifically for pattern recognition: an example is the WISARD machine designed at Brunel University and now produced commercially for robot vision.

Types of computer vision

The main classes of machine vision as it applies to robotics are as follows:

1) two-dimensional, isolated objects, binary image;
2) two-dimensional, isolated objects, grey scale image;
3) two-dimensional, with touching or overlapping objects;
4) two-dimensional inspection, e.g. checking components on printed circuit boards;

5) two-dimensional line tracking;
6) extraction of three-dimensional information from an isolated object using perspective, stereo, structured illumination or range finding;
7) as (6) but for a jumbled heap of objects;
8) three-dimensional scene analysis for mobile robot navigation, route finding and obstacle avoidance.

The last two classes, and indeed the others in some cases, are still in the research stage, whereas equipment is commercially available for the first six.

TWO-DIMENSIONAL VISION WITH ISOLATED OBJECTS AND A BINARY IMAGE

The feasibility of this is well established. The aim is to determine the position and orientation of an object, usually a fairly flat component such as a circlip or a leaf spring, as it passes along a conveyor, so that a robot can pick it up. The parts are usually all of the same type, so recognition is unnecessary. Many methods of processing the image exist: an example is shown in Figure 5.22, starting with a binary image (i.e. every pixel is black or white, with no intermediate grey levels). Such an image can be obtained from a television signal by thresholding it, i.e. assigning all brightness levels above a certain value to white and all the rest to black.

The well-known CONSIGHT system of General Motors uses special illumination to distinguish objects from their background. Two sloping sheets of light converge at the conveyor belt, forming a line of light. The field of view of a linescan camera arranged as in Figure 5.20 coincides with this line. The passage of an object along the belt causes the line of light to be displaced out of the camera's field of view; the result is a negative (dark on bright) image of the object.

Statistical pattern recognition

If the parts are not all of the same type, or if one type can lie in several stable attitudes (if it is nearly a cube, for example) then it is necessary to recognize the image as well as find position and orientation. A common technique for this is *statistical pattern recognition*. It is often computationally fairly cheap to extract several properties of the image of an object, such as its area, perimeter, aspect ratio, connectivity and maximum and minimum distances from centre to boundary. A set of values for these 'features' can be used as a signature for recognizing which of several possible objects is being seen. In an initial training session a signature is stored for each expected object; then when the system is run it compares the set of features measured on the image to be recognized with each stored signature using some statistical classification test.

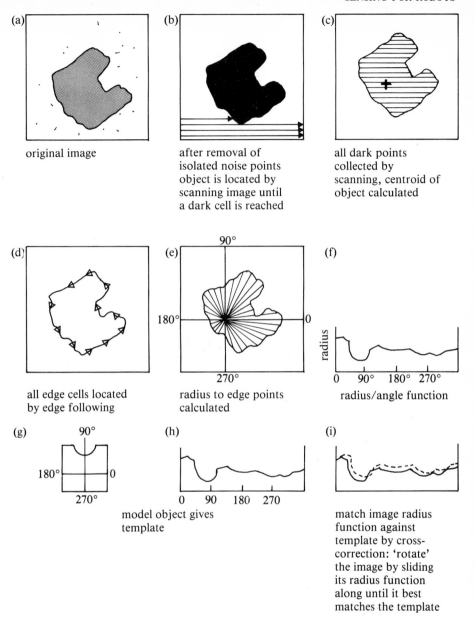

(a) original image

(b) after removal of isolated noise points object is located by scanning image until a dark cell is reached

(c) all dark points collected by scanning, centroid of object calculated

(d) all edge cells located by edge following

(e) radius to edge points calculated

(f) radius/angle function

(g) model object gives template

(h)

(i) match image radius function against template by cross-correction: 'rotate' the image by sliding its radius function along until it best matches the template

Figure 5.22 *An example of two-dimensional binary processing to determine the position and orientation of a flat, isolated workpiece.*

TWO-DIMENSIONAL VISION WITH ISOLATED OBJECTS AND A GREY SCALE IMAGE

If high contrast and even illumination cannot be guaranteed, the single brightness threshold used to distinguish dark areas from bright ones ceases to be adequate. By applying more complex processing to a grey scale image the boundaries of objects can be found even when the brightness change across the boundary is less than the brightness variation within a unitary region.

TOUCHING OR OVERLAPPING OBJECTS

If two or more objects overlap, a closed region in the image cannot be assumed to represent a single object, so it becomes necessary to recognize parts of objects and to work out how many objects make up the region and how they are disposed. The basic method starts by looking for features which can be recognized in the image. The features and the distances and angles between them are matched against a model of the object (Figure 5.23). In this example the features are circles of radius R_1 and R_2. The image analysis program attempts to fit circles of these radii to edges found in the image by an earlier stage of processing; two large and two small probable circles are found. (In this example we assume that the objects are essentially flat; if this is not so the problem may be harder, since, if one object leans on another, the projection of, say, a circular feature will be elliptical. The problem then starts to be one of three-dimensional scene analysis.) The model (Figure 5.23(c)) states that there must be one large and one

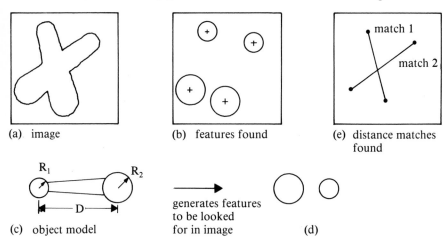

(a) image (b) features found (e) distance matches found

(c) object model generates features to be looked for in image (d)

Figure 5.23 *Model-based vision in two dimensions.*

small circle, separated by D; by examining all possible pairs of a large circle with a small one, only two distance matches are found, and it is possible to deduce that the image is of two of these objects lying across each other. This is a primitive example of model-based vision.

TWO-DIMENSIONAL INSPECTION

Machines are available for checking whether printed circuit boards have broken tracks, missing holes, incorrect components and so on. The crudest approach is to store an image of a perfect board and to compare it pixel by pixel with the image of the board being tested. This is of limited use as it relies on perfect registration between the two, and is just as likely to pick up some irrelevant difference as a genuine fault. More sophisticated systems keep a list of regions to be checked and within each region carry out a model-based check, looking for key features. Such systems can be used for inspecting other parts and assemblies, as long as the area to be inspected is fairly flat.

TWO-DIMENSIONAL LINE TRACKING

In welding or gluing a seam or cutting material by following a pattern, a robot must track a dark or light (or more complex) line. Figure 5.24 shows the principle in a simplified form. The camera may be attached to the arm. Once the system is locked onto the seam a narrow strip lying across it is repeatedly scanned, and from the brightness profile of the strip the seam position is calculated.

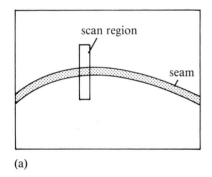

(a)

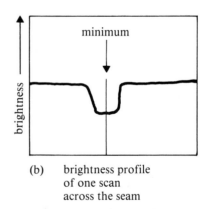

(b)   brightness profile
of one scan
across the seam

Figure 5.24 *Visual seam tracking.*

THREE-DIMENSIONAL VISION: ISOLATED OBJECTS

This is the problem of looking at, say, a casting dropped at random on a conveyor and determining its orientation in three dimensions. Another example is looking at a weld seam between two plates which are not accurately located to determine exactly where to put the welding gun.

The vision system may be designed to measure some specific dimensions of a restricted class of object; this is so with the welding example. Here the light-stripe technique of Figure 5.21 can be used. Processing is complicated by the extreme noisiness of the images because of light from the arc and splashes of molten metal.

The example of the casting on the conveyor needs a more general vision system. There are four main methods of obtaining three-dimensional information in such a case.

Single image (monocular vision)

A single image is in theory infinitely ambiguous, but in practice it is usually known that only a few interpretations make sense; e.g. the distance from the object to the camera is known, so there is no doubt about its size. A model-based approach is used, with a three-dimensional model. It is made up of cylinders, cones, blocks and so on. By calculations of the sort used in computer graphics, the appearance of the model from a given direction can be found (Figure 5.25). This generates a two-dimensional model which can be compared with an image as described earlier.

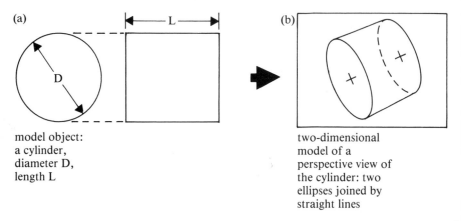

(a) model object: a cylinder, diameter D, length L

(b) two-dimensional model of a perspective view of the cylinder: two ellipses joined by straight lines

Figure 5.25 *Model-based vision in three dimensions.*

Stereo vision

If a pair of cameras is used, or a single one moves relative to the target object, depth information can be extracted by triangulation. The method depends on the ability to match small regions of the scene in the two images; this requires a lot of computing and is usually too slow for robotic purposes. In theory stereo vision works for an arbitrary scene, but when it comes to describing what is seen is a useful way geometrical models are again resorted to.

Structured illumination

This method, already described for a simple case, is essentially a method of extracting range from one image by triangulation, and so in principle is similar to stereo vision. Its main advantage is that processing is easier, since it is not necessary to match two scenes but merely to pick out the pattern made by the illuminating fan-beams as they lie across the object.

Range imaging

Structured illumination and stereo vision are special cases of rangefinding. More direct methods of rangefinding are possible, in which a scene is scanned, sometimes mechanically, by a beam of light or ultrasound, the distance to a reflecting object being measured by the time of flight of a pulse or by phase methods. Direct rangefinding is best suited to outdoor scene analysis or vehicle navigation.

THREE-DIMENSIONAL VISION APPLIED TO A HEAP OF PARTS

This problem is important because in many factories parts are transported jumbled up in bins or boxes. Small components can be orientated by vibratory feeders, but this is not possible for larger items.

The basic approach is to extend one of the three-dimensional isolated object methods by model-based matching. It is not yet possible to do this reliably in most cases.

THREE-DIMENSIONAL SCENE ANALYSIS FOR MOBILE ROBOTS

The problem is rather similar to that of the heap of parts and is a subject of much current research.

Non-visual sensing in welding and other processes

There are some processes in which the movements of the industrial robot can be determined by signals monitoring some aspect of the process itself. The most prominent example is arc welding. The distance of the electrode

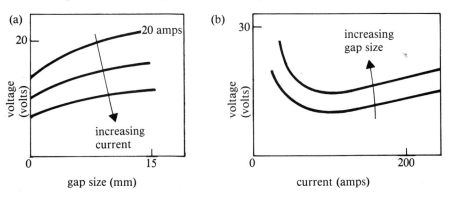

Figure 5.26 *The relationships between current, voltage and arc gap in arc welding.*

from the workpiece can be estimated from the arc current and voltage; as shown in Figure 5.26 there is a (non-linear) relationship between these quantities. The need to control the arc gap is independent of the need to track the seam, so this current–voltage method may or may not be associated with a vision system.

Bibliographic notes

The most detailed sources of information on transducers are manufacturers' catalogues. Sensors for robotics are developing rapidly and these developments are reported in conference proceedings and journals such as *Sensor Review*. A recent text on transducers is Alloca and Stuart (1984).

Piezoelectric polymer touch sensors are described in Bardelli *et al.* (1983) and Dario *et al.* (1983). An account of the optical imaging touch sensor is given in Mott *et al.* (1984), and of a related device using an array of phototransistors instead of a CCD camera in Tanie *et al.* (1984). Experiments on thermal touch sensing are reported in Russell (1985).

As remarked in the text, computer vision is a vast subject to which justice cannot be done here. An intelligible text on the fundamental mathematics of image processing is Rosenfeld and Kak (second edition, 1982). A survey of model-based vision systems is given in Binford (1982). A paper explaining one structured-illumination vision system for arc welding in detail is Clocksin *et al.* (1985).

Chapter 6

Performance Specifications of Industrial Robots

This chapter deals with the physical characteristics of robots; the economic aspects of performance, such as reliability, are covered in Chapter 12.

Physical specifications have several uses: choosing a robot for a given task; assessing whether a robot's performance has degraded with time; planning a task so that it can be done by a given robot; as targets for the design of new robots; and as a basis for designing end effectors, including devices for enhancing the performance of the basic robot.

Because of the great variety of shapes and uses of industrial robots, standardization of specifications over all robots is difficult. However, there are certain characteristics which, all else being equal, allow robots of similar type to be compared, and these are listed in the following sections. An international standard is being prepared by the International Standards Organization (ISO). It introduces a number of conventions such as a three-way division of a robot into a group of major primary axes (the arm), a group of secondary minor axes (the wrist) and the end effector, with a clearly defined mechanical interface between the wrist and the end effector. Co-ordinate systems are defined for the arm and the mechanical interface (Figure 6.1).

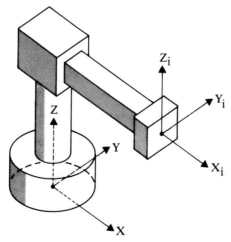

Figure 6.1 *Coordinate systems of the ISO standard.*

In Britain the Machine Tool Industry Research Association (MTIRA) is developing a set of standard test procedures which can be carried out on site. The performance parameters to be tested have not been finalized, but a tentative list is as follows:

1) the load-carrying capacity,
2) repeatability of positioning,
3) path tracking ability,
4) consistency of velocity,
5) positioning time,
6) static and dynamic stiffness characteristics,
7) vibrational behaviour.

Geometric configuration; number of axes

The subject of arm geometries and degrees of freedom has been dealt with in Chapter 2 and will not be discussed further except to say that manufacturers of the cheaper robots tend to gloss over the question of the number of axes. A truly general purpose robot needs at least six controlled degrees of freedom, excluding the gripper, but, since three-axis wrists are much more difficult or expensive to make than those with two axes, cheap robots sometimes leave the third axis off.

Assembly robots often use a simple pneumatic cylinder for one axis, with servo control of the others.

Positioning accuracy and repeatability

ACCURACY

The accuracy with which a robot can bring the payload to a position and hold it there or the accuracy with which it passes through a position while moving, can both be important. Perhaps because of the difficulty of measuring the second of these, accuracy is usually defined for the static case, when the manipulator has approached a target point and is holding the payload in a fixed position. Since this is done by servo control (expect for pick and place machines) and servos are never perfect, there will be both an offset and a random error. This is true for each axis, and the size of the error will not be the same for all axes. If a single figure is quoted for a guaranteed maximum position error for the whole robot it should be the worst case; the accuracy in certain axes may be much better.

Accuracy is also a function of the geometry and load at the time: the robot will tend to deflect under heavy loads and the increased inertia may affect the servos; and geometry affects accuracy in that often what is controlled is joint angle, so that when the arm is extended the positional error is greater (Figure 6.2). Note that since prismatic joints are usually more rigid than revolute ones, the most accurate manipulators are Cartesian, those with some prismatic joints and some revolute are intermediate, and those with all revolute joints are the least accurate, in principle at least.

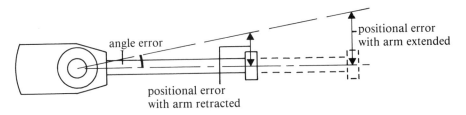

Figure 6.2 *The effect of arm radius on positional accuracy.*

There may also be a trade-off between accuracy and speed: if more time is allowed for the servo to settle down to a commanded position, higher accuracy may be obtained, at the cost of a lower overall speed.

REPEATABILITY

Accuracy as just discussed is a measure of how closely the robot approaches its target, on average. Repeatability is a measure of how closely the achieved position clusters around its mean. The difference between accuracy and repeatability is illustrated by Figure 6.3. Repeatability is often more important than accuracy since, provided that the accuracy error is constant, it can be allowed for. (This is only true if the robot keeps repeating the same cycle of actions.)

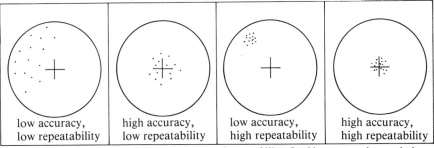

Figure 6.3 *The difference between accuracy and repeatability. In this target analogy each dot represents an attempt to get to the central cross. The size of the cluster shows the spread in the result, and the closeness of the centre of the cluster to the cross is a measure of the accuracy.*

Accuracy and repeatability are usually of the same order, typically millimetres for very large robots, tenths of a millimetre for general purpose robots and hundredths of a millimetre for the most accurate assembly robots.

The ISO standard will define several accuracy parameters, such as

1) local pose accuracy,
2) one-way pose repeatability,
3) multiway pose repeatability,
4) stability, stabilization time,
5) overshoot on reaching a target point,
6) path-following accuracy,
7) path repeatability,
8) velocity fluctuation,
9) overshoot and undershoot on the transition between two straight paths.

It also distinguishes between (a) the desired pose (pose means position and orientation of the payload or of some reference point on the robot such as the end effector mechanical interface), (b) the programmed pose, which is the robot's stored estimate of the desired pose, (c) the commanded pose, which is the control unit's intrepretation of the commanded pose, and (d) the pose actually attained. Errors can arise at any stage of the chain from desired to attained pose. (It may not be possible to test all these.)

TEST METHODS FOR ACCURACY AND REPEATABILITY

To be fully useful, specifications should be measured in a standard way. Such standards are still being defined, although each manufacturer has standard tests for its own robots. Among the methods in existence may be mentioned the use of a matching cube, held by the robot, and a fixed corner (sometimes called a trihedron). Each face of the corner is fitted with high resolution distance sensors such as LVDTs so that, when the robot tries to fit the cube into the corner by moving the cube to the calculated position of the corner, any positional and angular errors can be measured.

This method measures the errors in a static condition, or when the robot brings a payload to a target point and stops. Measuring its performance while it is moving over larger distances is harder. The basic method is to track a marker carried by the robot, using multiple television cameras or triangulation with several rangefinders. It is not easy to achieve high accuracy of measurement over a large volume.

Angular accuracy and repeatability

The angular accuracy of any revolute *arm* joint is the determinant of the positional accuracy for that axis. The angular accuracy of a *wrist* joint determines the accuracy with which the payload is orientated. The same applies to repeatability. Therefore, a robot specification should include the angular accuracy and repeatability of all the wrist joints.

Speed

A manufacturer's specifications will include speed, but this is often the maximum steady speed with the arm fully extended. In practice the arm has to accelerate and decelerate so its average speed is lower than the maximum, particularly for short strokes. Also, it will often not be fully extended and the speed will again be lower.

For a robot with continuous path control the speed when slewing in an unconstrained way from one point to another may be much higher than that with which it can follow a prescribed path.

Maximum and minimum acceleration are not usually quoted. In applications needing low accelerations so as not to spill liquids or break fragile objects continuous path robots can be used and taught or programmed with a trajectory having no sudden accelerations.

The maximum acceleration is of interest more because it governs the effective speed than for its effect on the payload, although operations are conceivable, such as shaking a workpiece, where high acceleration is important.

The ISO standard will include:

1) individual axis velocity (maximum rated),
2) resultant velocity (maximum rated),
3) maximum path velocity under continuous path control, at some specified accuracy,
4) acceleration under various conditions (axis, resultant, path),
5) minimum positioning time at rated load, for a specified travel distance and path accuracy.

Speed and acceleration accuracy

Applications needing good speed control, such as the application of a line of sealant, are less common than those needing quality of position control only, but are not unknown. Such tasks are generally done by continuous path robots, in which the control of speed and the control of position are intimately related.

Spatial specifications: working volume, swept area, reach

The next group of specifications is concerned with the space which can be accessed by the robot. This can be expressed as the volume accessible to the payload and this number can be used for comparing robots, but other measurements are important as well. This aspect of robot performance is highly dependent on the arm configuration. The most usual way of describing the working volume is as a plan view and side elevation of the area swept out by a joint or combination of joints, as shown in Figure 6.4. Such a representation cannot capture the full three-dimensional shape:

both plan and elevation have to be drawn for a fixed state of some joints. These drawings can give a misleading impression; e.g. for a spherical polar arm geometry the plan of the swept area is correct when the gripper is level with the shoulder, but at the top and bottom of the arm's reach the swept area is a narrow ring.

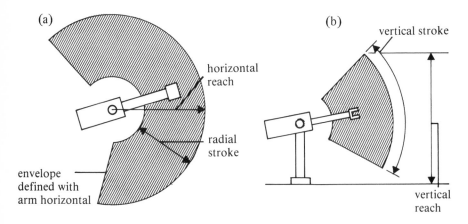

Figure 6.4 *The work envelope of a spherical polar robot in (a) plan and (b) elevation.*

Two terms characterizing the working envelope are *reach* and *stroke*. These are not used consistently, but reach is essentially a distance from a reference point (the robot's shoulder for horizontal reach, or the floor for vertical reach) whereas stroke is the range which can actually be moved through. Both reach and stroke apply to the end effector.

Payload (maximum load capacity)

In this context the payload should refer to a workpiece or tool and does not include the gripper, which is regarded as part of the robot. However, some manufacturers supply an arm ending in a mounting to which various grippers can be attached, and in this case the weight of the gripper must be subtracted from the payload capacity figure given. If the robot is dedicated to one purpose and has a tool built into it then a payload figure is not applicable.

Since load affects speed and accuracy, it may be quoted for more than one condition. Load capacities range from less than a kilogram to several tons.

A rigorous specification of load needs to include several parameters rather than just a single payload figure. These are the allowed mass and moment of the load (referred to the end effector interface) for which the specified performance is met (Figure 6.5) and the maximum mass, torque and applied thrust which will not damage the robot.

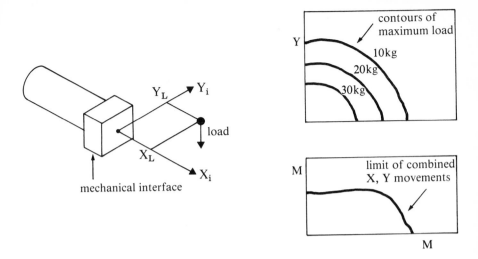

Figure 6.5 *Contours of allowed load and moment defined in wrist/end-effector interface coordinates.*

Control-related specifications

MEMORY CAPACITY

For a limited sequence (pick and place) arm or a point to point robot, memory capacity is expressed as the number of movements or positions, and may be several hundred. Such a number might well be needed in, say, spot welding; for many transfer operations less than ten positions might be used. For a continuous path robot, memory capacity is expressed as the length of time which can be recorded.

The specification should state what kind and capacity of exchangeable memory device is provided.

PROGRAM STRUCTURE

These remarks about memory capacity assume that a program consists of a simple sequence of operations. More complex facilities are useful, such as subroutines, branches, a choice of programs and so on. The details of these facilities should be stated.

ADVANCED FEATURES

Some of the properties whose presence or absence, and their type when present, should be made clear in the specification are as follows:

1) programming languages,

2) ability to generate circles,
3) ability to generate welding patterns ('weaving'),
4) interfaces for sensors (such as vision systems),
5) ability to track a conveyor,
6) ability to control ancillary devices such as positioning tables,
7) communications ports for factory networks,
8) ability to be down-line loaded with a program by some other system.

Vibration

Several vibrational parameters can be specified, such as

1) resonant frequencies of the robot structure,
2) amplitude of vibrations produced by the robot (some vibration is always present in some servo systems),
3) frequency response to applied vibration,
4) damping,
5) dynamic stiffness,
6) resistance to external vibration.

Miscellaneous specifications

1) Stiffness
2) danger volume (swept volume including all moving parts)
3) mounting positions allowed
4) fixing methods
5) transport methods
6) weight of each part
7) cables, hoses, accessories
8) power supplies needed (electric, hydraulic, pneumatic)

In addition there are specifications concerning safety, reliability and environmental conditions; these are discussed in Chapter 12.

Bibliographic notes

The draft standard from the International Standards Organization is ISO/TC184/SC2/WG2N6. Information on MTIRA's activities in standardization can be obtained from MTIRA, Hulley Road, Macclesfield, Cheshire SK10 2NE, England.

Chapter 7

Applications of Industrial Robots

In 1980 when Engelberger published his book *Robotics in Practice* on the technology and applications of robots, he found them used in the 15 industrial processes listed below:

1) die casting
2) spot welding
3) arc welding
4) investment casting
5) injection moulding
6) forging
7) press work
8) spray painting
9) foundry work
10) machine tool loading
11) heat treatment
12) deburring
13) palletizing
14) brick manufacture
15) glass manufacture

He was also able to report applications not then in regular use but under development, such as packaging, electric harness manufacture, assembly and sheep shearing. Most of these can now be found in production use.

Since 1980 there has been explosive growth in the range of robot applications, and it is impossible to list them all. Therefore this chapter will focus on a selection of applications of particular interest, generally because of their widespread use or because they point the way to the future.

It should be noted that although it concentrates on the details of a task done by an individual robot, and increasing numbers of isolated robots will continue to be installed, there is a trend towards integrating robots into manufacturing systems along with other machine tools. This implies that in future the emphasis in robotics will be on the relationship of the robot to the rest of the system rather than on how an individual robot works.

The well-established applications will be discussed first, and then some still in the research stage. The chapter concludes with a section on the integration of robots into the workplace.

Machine loading

The first application of industrial robots was in unloading die-casting machines. In die casting the two halves of a mould or die are held together in a press while molten metal, typically zinc or aluminium, is injected under pressure. The die is cooled by water; when the metal has solidified the press opens and a robot extracts the casting and dips it in a quench tank to cool it further. The robot then places the casting in a trim press where the unwanted parts are cut off. A robot serving two die-casting machines and a trim press is shown in Figure 7.1. The robot often grips the casting by the sprue. (The *sprue* is the part of the casting which has solidified in the channels through which molten metal is pumped to the casting proper. Several castings may be made at once; in this case they are connected to the sprue by *runners*.) When the sprue and runners are cut off by the trim press, the press must automatically eject the casting(s) onto a conveyor.

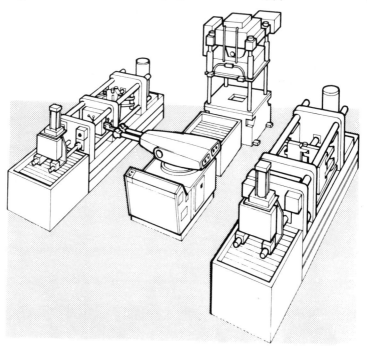

Figure 7.1 *A robot serving die-casting machines.*

Many combinations of robot, die-casting machines and presses are possible. In some cases the robot fits inserts into the die before casting takes place. It may have to spray the die with lubricant. A point to point robot can be used. Its geometry must allow it to reach into the presses. With proper placement of the machines a two-axis wrist may be sufficient. The cycle time of the die-casting process is at least several seconds, so a relatively slow robot can be used. Interlocking with sensors and microswitches is needed to ensure that the robot grasps and removes the casting successfully, that the trim press is emptied correctly and that the die-casting machine and trim press are open when the robot tries to insert its arm.

Robots are now used for loading and unloading injection moulding machines (some of which make plastic items as large as dustbins and boat hulls), lathes, milling machines, sheet metal presses and other machines. Figure 7.2 shows a Cincinnati Milacron hydraulic robot loading machines in a manufacturing cell consisting of a computer numerical control lathe, a gauging station and an induction hardening machine. Figure 7.3 shows a machine tool with a loading robot built in.

Figure 7.2 *A robot transferring parts betwen the machines of a manufacturing cell. It is shown loading a turned part into a gauging station (courtesy of Cincinnati Milacron).*

Figure 7.3 *A numerically controlled lathe with a built-in loading robot (courtesy of Cincinnati Milacron).*

Machine loading is straightforward in principle but needs attention to detail, as may be seen at exhibitions where, largely because the setting up has been done in a hurry, robots can often be seen just failing to grasp workpieces properly and then loading non-existent workpieces into machine tools. This illustrates, negatively, what is meant by machine intelligence.

Pallet loading and unloading

Stacking boxes, sacks of cement or bricks on a fork-lift pallet in a stable and space-efficient way requires the robot to put each item in a different place (Figure 7.4). A point to point robot can be taught or programmed all these locations individually. A step-and-repeat facility is sometimes provided so that as long as the stack is regular it is necessary only to teach the first position, the increment and the number of steps in each direction.

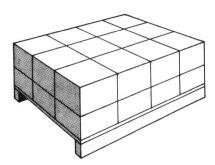

Figure 7.4 *Arrangement of boxes on a pallet: each must be placed in a different position, and often there is more than one orientation needed to pack the maximum number of boxes in.*

A practical problem with palletizing is the gripping of the objects in such a way that the gripper does not collide with the items already in place. Special grippers with thin blade-like jaws can be designed; also, instead of dropping each object straight down it can be brought in with a slight sideways movement; another way is to grip the object by its top surface with a suction gripper.

Figure 7.5 shows cylinder heads being palletized by a KUKA IR 160/60. The large black box contains a vision system which determines the type, position and orientation of each cylinder head. These come in two sizes, and are placed on two separate pallets according to size.

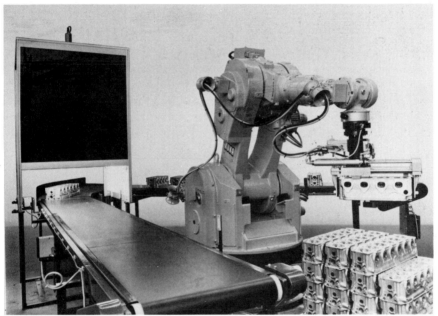

Figure 7.5 *A KUKA IR 160/60 palletizing cylinder heads (courtesy of KUKA Welding Systems & Robots Ltd).*

Investment casting

This uses the lost-wax process: a wax pattern is made on which a mould is formed by repeatedly dipping the pattern in a ceramic slurry until a thick coat is built up; the ceramic is fired in a kiln (after the wax has been melted and run out); metal is cast in the ceramic mould; and finally the mould is broken to release the casting. The stage of repeatedly dipping the pattern in the slurry has been done by robot. The only unusual feature is that the mould has to be spun to spread the slurry evenly over the surface and throw off any excess. This is done by a special gripper incorporating a spin motor.

Spot welding

The spot welding of car bodies is the most well-known use of industrial robots, mainly because the motor industry is in the public eye more than most; also, a spot welding line with its showers of sparks and large number of robots is more spectacular than a solitary robot unloading a die-casting machine. A typical spot welding line is shown in Figure 7.6. Twelve Cincinnati Milacron HT^3 robots on each line together make 300 spot welds on each body shell; this line can handle 43 bodies an hour.

Figure 7.6 *A car body spot welding line — Ford Sierras being welded by Cincinnati Milacron HT^3 robots (courtesy Cincinnati Milacron).*

In spot welding, which is only useful for moderately resistive metals such as mild steel, in thin sheets, two pieces to be joined are clamped between copper electrodes and enough current is passed through the point of contact to heat the steel to melting point by resistance heating. AC power can be used, and is applied for typically 1/5 s. The welding head needs thick cables and cooling hoses and can weigh as much as 100 kg (unless supported, as manually operated welding guns are), so a powerful robot is needed. Most of the time is spent moving between welds, so it needs to be fairly fast. Accuracy need not be extreme: a few millimetres may be good enough.

Spot welding is done both on static assemblies on indexed conveyors or AGVs (such as Fiat's Robogate system) and on assemblies on continuously moving lines; in this case the robot must be capable of being taught on a static assembly and then tracking a moving one. It will be able to make only a limited number of welds before the assembly moves out of the tracking window (see the last section of this chapter). The production line of Figure 7.6 uses a discontinuous conveying method.

Arc welding

Arc welding as it applies to robotics generally uses the metal–inert gas (MIG) technique shown in Figure 7.7. An arc is struck between the workpiece and a wire of filler metal which is slowly extended as it is consumed. It is surrounded by a tube through which argon or helium is blown to protect

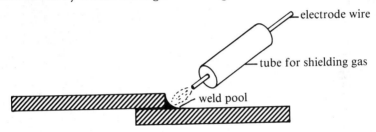

Figure 7.7 *The basic arrangement for arc welding.*

the weld area from the oxygen and nitrogen of the atmosphere, both of which can combine with the weld metal at the temperature of an arc. Arc welding can be used with a wide variety of metals including aluminium alloy and stainless steel. Typical arc conditions are 100 A at 20 V and a gap of several millimetres.

In arc welding the electrode is often oscillated at right angles to the seam or moved in small circles; this is called weaving and is needed to make the weld of adequate width. Also, several passes may be made along the seam to build up the thickness of the weld. Therefore welding robots have facilities for weaving and multiple passes. These can be control functions, or weaving can be done by an oscillating welding torch. Circular joints must often be welded, so software for this is also provided. An alternative is to mount the workpiece on a rotary positioning table and to rotate it while welding. A robot welding a circular seam is shown in Figure 7.8, and another arc welding robot is shown in Figure 7.9.

As explained in Chapter 5, arc welding benefits from sensory feedback, both for control of the voltage and arc gap and for following a seam.

Arc welding obviously needs a robot with continuous path control.

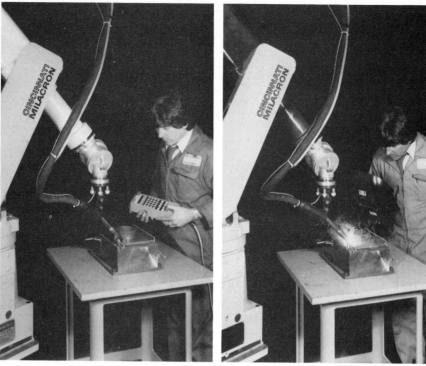

Figure 7.8 *A Cincinnati Milacron T³-726 arc welding a circular seam (courtesy Cincinnati Milacron).*

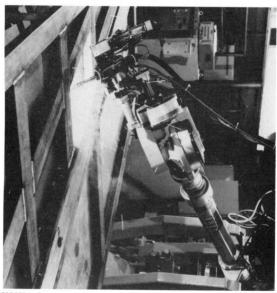

Figure 7.9 *A KUKA IR 662/100 arc welding a straight seam (courtesy of KUKA Welding Systems & Robots Ltd).*

(a)

(b)

Figure 7.10 *Paint spraying by a Spine robot: (a) a van interior; (b) a steel cabinet (courtesy Spine Robotics).*

Spraying (paint, enamel, epoxy resin and other coatings)

Because many pigments and solvents are poisonous, the automation of paint and other types of spraying is desirable for health reasons as well as for reasons of economy and consistency. Continuous path robots are needed, but need not be very precise. Since the solvent-laden atmosphere is potentially explosive, precautions have to be taken to avoid sparks. The workpieces often move on a continuous conveyor, so the ability to program or teach on a stationary workpiece and then to reproduce the action while tracking a moving one is commonly needed.

The number of degrees of freedom needed depends on the workpiece: for spraying flat panels on one side a rather simple robot can be used, whereas to reach into the interior of a car body shell requires a more dextrous one, such as the Spine robot (Figure 7.10), whose unusual geometry allows it to reach like a snake or elephant's trunk through small openings.

The successful use of robots for spraying depends very much on care of the spraying equipment to avoid clogging, mixing of colours and so on.

Spraying robots are generally taught by lead-through or walk-through, sometimes using a teaching arm. When the stored program is replayed the actual movements can be modified by feedback from a conveyor so as to track a moving workpiece.

Fettling (grinding, chiselling); polishing

In fettling and polishing the robot either applies a power tool to a fixed workpiece or holds the workpiece and presses it against the tool; which of these is done is dictated by the size and weight of the tool and workpiece. A grinder may be powered by a motor of 20 kW or more, which is too heavy for most robots. Figure 2.13 shows a robot holding a casting aginst an abrasive wheel to cut off the sprue. In Figure 7.11(a) the remaining stub is ground off on an abrasive belt. In a third operation the hollow underside of the casting is presented to a pneumatic chisel which removes any sand left from the casting process (see Figure 7.11(b)).

If an object is to be polished all over by holding it against a wheel or belt, it will have to be put down in a jig and regrasped so that the part by which it was originally held can be polished; this might be termed the Achilles' heel problem.

In grinding it is often necessary to follow the edge of, say, a casting. Such an edge-following operation is shown in Figure 7.12. The amount of metal to be removed can vary along the edge; one way of coping with this is to mount the robot-held tool in a resilient suspension which exerts a constant grinding force, with the tool free to move until it comes up against a stop. The robot is programmed with a path which follows the edge of the casting, and must move slowly enough for the tool to grind its way to the stop at every point along the path.

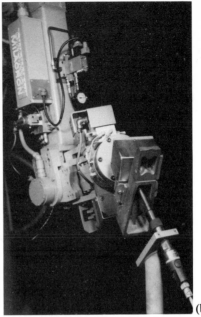

(a)   (b)

Figure 7.11 *A Cincinnati Milacron T³–586 (a) grinding flash off castings following the cutting off of the sprue shown in Figure 2.13, then (b) removing sand by presenting the casting to a pneumatic chisel (courtesy Cincinnati Milacron).*

Fettling by robot is important because it is an unpleasant and unhealthy job: dirty, noisy and dangerous.

Cutting

Robots are used for flame, laser, plasma torch and water jet cutting. Figure 7.13 shows a robot being used for water jet cutting of motorcycle helmets, and Figure 7.14 shows a robot using a plasma torch.

Inspection

Robots can be used for several kinds of inspection. One is to establish the dimensions of an object by probing with a touch probe; an example of this is described later. Another is presenting a part to a camera for visual inspection. A third is ultrasonic inspection. In Figure 7.25 a robot scans the surface of a laminated panel with an ultrasonic probe under water.

Training and education; hobby robots

There are several categories of education and training in which there can be a need for robots:

1) training the workers who will use the robots in a specific factory;
2) teaching robotics as part of a certificate, diploma or degree course in engineering;
3) using a robot as a versatile and interesting peripheral when teaching computing in schools and adult education.

The training of production and maintenance workers must be based on the type of robot to be used, although it may be more convenient and cheaper, and give a broader understanding of robotics, to use other types as well.

The growth of the other two categories of education has led to the development of educational robots costing hundreds or thousands of pounds, instead of the tens of thousands typical of industrial robots. These cheap robots are also aimed at the hobby market, or sometimes at relatively undemanding industrial applications. Educational and/or hobby robots sacrifice performance for cheapness. They are generally powered by small stepper

Figure 7.12 *A contouring (edge-following) operation with a robot held grinder (courtesy KUKA Welding Systems & Robots Ltd).*

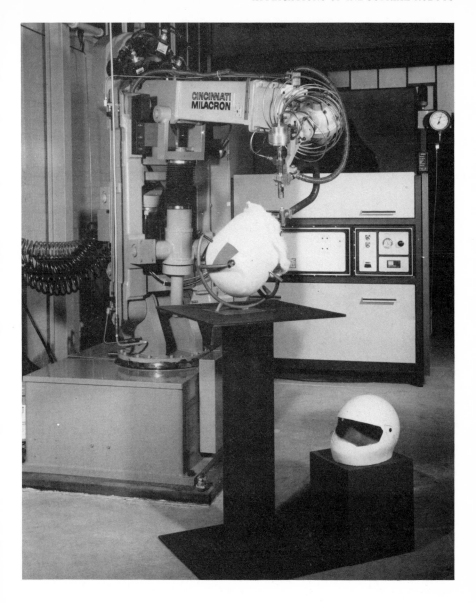

Figure 7.13 *Cutting to shape of glass reinforced plastic crash helmets with a high pressure (3000 bar, 50,000 psi) water jet. The process produces a clean surface without creating injurious dust (courtesy of Cincinnati Milacron).*

motors, which makes them fairly slow and of limited load capacity (a few hundred grams) but allows simple interfacing and software for the personal computers used to control them. They are usually small, with a vertical and horizontal reach of perhaps 0.5 m, to reduce the torque on the motors and to increase positional accuracy. Their accuracy is poorer than that of an industrial robot of the same size. They often have only five degrees of freedom (plus the gripper movement). An example is shown in Figure 7.15.

The use of robots for teaching subjects other than robotics is an interesting development; it is seen with mobile robots ('turtles') as well as with arms.

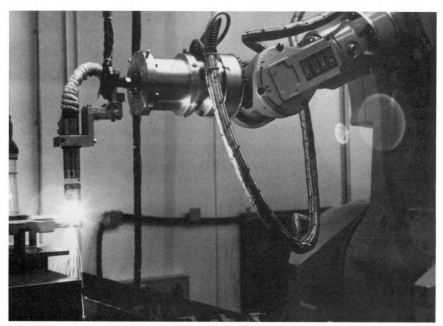

Figure 7.14 *A KUKA IR 160/60 robot being used for plasma cutting (courtesy of KUKA Welding Systems & Robots Ltd).*

The attraction of a robot is that many concepts in computing and electronics such as subroutines, coordinate transformation and analogue-to-digital conversion can be made vivid, and many enjoyable but instructive projects, problems and demonstrations can be devised.

Robots in assembly

The application of industrial robots in assembly has perhaps the greatest potential for growth. The automation of assembly has lagged behind that of primary component manufacture because it needs much dextrous manipulation. Now many companies perceive that robotic assembly

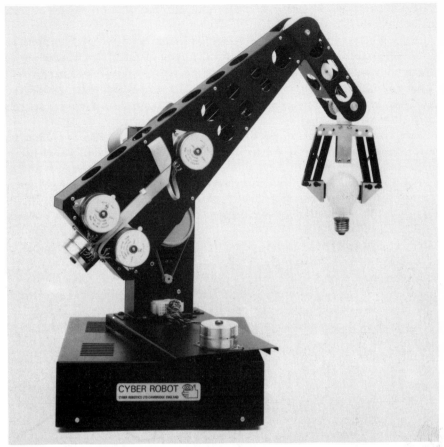

Figure 7.15 *An educational robot, the Cyber 310, it has a two axis wrist (courtesy of Cyber Robotics Ltd).*

automation can reduce costs and are in a race to develop the next genera-tion of factories. To illustrate the scale of this change, IBM is spending $350 million on modernizing a single typewriter factory.

Of course, automatic assembly is not new: it has been done by hard automation (i.e. using machines dedicated to the assembly of a single pro-duct) for many years, but hard automation is not appropriate for many products, e.g. if production runs are short. Production runs of some pro-ducts are shortening because the products become obsolete more quickly than formerly, and because they are made in many varieties.

TYPES OF ROBOT FOR ASSEMBLY

Almost any configuration and control method can be and has been used, but the field is dominated by a few types. Pick and place arms, usually of cylindrical polar geometry, are often used as they are fast, relatively cheap and of high precision; a repeatability of ±0.02 mm (about ±0.001 in) is typical.

Among servo-controlled robots, the Cartesian and SCARA configurations dominate. Both of these have a mechanically defined vertical axis, whcih allows components to be inserted vertically without the straightness of the motion depending on servo quality. The cylindrical polar configuration also has this property, but is less common.

Despite not having a vertical prismatic joint, all-revolute jointed arms such as Unimation's PUMA are often used for assembly. The PUMA does have a good working envelope, but its popularity is probably due more to the fact that it was the first arm of its speed and precision to become readily available. Among the most recent assembly robots the SCARA configuration is perhaps the most common. An example is the IBM 7545 (see Figure 7.16). Repeatability is ±0.05 mm (±0.002 in). IBM also make a gantry–Cartesian assembly robot, the 7565, which, unusually for an assembly robot, is hydraulically powered using an actuator in which a group of four rams on the moving assembly in sequence press rollers against a wavy cam-bar (Figure 7.17). The position transducer is also unusual, based on the travel time of a magnetostrictively generated pulse along a wire. The wrist uses vane-type rotary hydraulic actuators.

Figure 7.18 shows the Pragma 3000, a Cartesian robot used for assembly. In addition to robots of conventional geometry, new types such as the parallel-structure Gadfly (Chapter 2) are intended for assembly, being designed for high speed and precision. The ASEA IRb 1000 is a polar robot

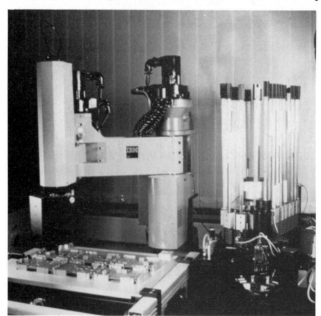

Figure 7.16 *The IBM 7545 robot (courtesy IBM United Kingdom Ltd).*

mounted, with the arm hanging vertically, on a gantry (Figure 7.19). It is very fast (up to 12 ms^{-1}), as are other recent assembly robots such as the Adept, an extremely fast SCARA robot.

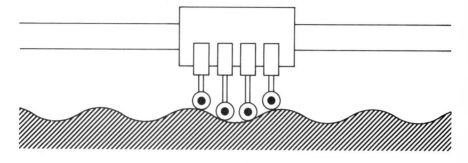

Figure 7.17 *The hydraulic actuator of the IBM 7565 gantry robot.*

Figure 7.18 *A pair of Pragma A3000s assembling cylinder heads; the robot in the foreground is fitting a valve spring (courtesy of Fairey Automation Ltd).*

GRIPPERS

Assembly imposes severe demands on gripper design, partly because of the high precision with which the workpiece must often be located relative to the partially completed assembly, and partly because components of several different shapes have to be handled. Sometimes there may be a robot for every component type; otherwise the robot must have several grippers on a turret, or be able to change grippers, or have a gripper which can adapt to different components. In the flexible manufacturing cell of

Figure 7.20, which uses two IBM 7545s, resistors, capacitors and diodes are inserted by robot 1 and integrated circuits by robot 2. Robot 1 uses two different grippers, one for axial lead components (resistors and diodes) and the other for capacitors. Each jaw of the permanently fixed gripper of robot 2 is divided into a stack of narrow fingers. By extending only the appropriate number of fingers it can match any standard dual-in-line package. If this were not done, then when a small integrated circuit, say an eight-pin one, was inserted with a gripper long enough to hold a 20-pin integrated circuit the jaws would collide with components already present.

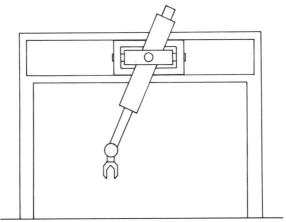

Figure 7.19 *The ASEA IRb 1000 polar robot.*

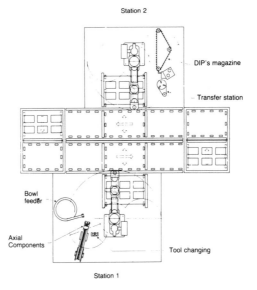

Figure 7.20 *A plan of the manufacturing cell for circuit board filling (courtesy IBM United Kingdom Ltd).*

Circuit board filling (stuffing) demands very high accuracy: there may be a tolerance of only ±0.1 mm when inserting a wire in a hole. The gripper jaws have grooves on their inner faces into which the component leads fit. These must be aligned with the holes in the board before the component is pushed down by an axial piston (Figure 7.21). The leads are bent to the correct angle and cropped by fixed tooling before being picked up by the robot.

Mechanical assembly, as opposed to circuit board stuffing, usually needs tools such as screwdrivers. The options for using different tools are the

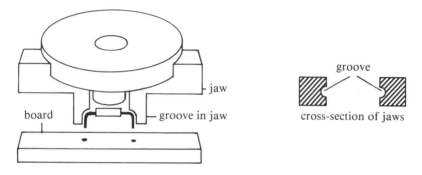

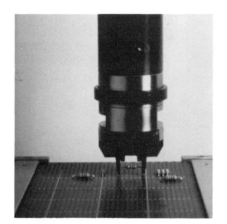

Figure 7.21 *A gripper design for insertion of resistors and diodes (courtesy IBM United Kingdom Ltd).*

same as those mentioned earlier for gripping a range of electronic parts. Tool changing from a magazine is relatively slow and presents interface problems; a multi-tool end effector with a turret is bulky and heavy; using one robot for each tool implies a large and expensive installation. All three methods have their place. Among the tools used are autoscrewdrivers in which each screw (often self-tapping) is fed pneumatically through a tube to a guide below the blade, sealant dispensers and nut-runners.

COMPLIANCE

As remarked in Chapter 2, assembly is where compliance is needed, particularly for inserting shafts into holes and fitting gears, bearings and so on onto shafts. The RCC device is one solution to this kind of problem but does not work for the sort of misalignment shown in Figure 7.22 as there is no mechanical feedback indicating the direction in which to move. The only solutions are to search blindly for the hole by moving the component in some pattern until it slides in, or to use visual sensing. For blind search the selective compliance of a robot with a vertical axis allows decoupling between the horizontal searching movement and the vertical insertion stroke.

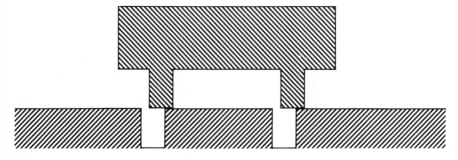

Figure 7.22 *Misalignment problem needing active compliance.*

DESIGN FOR ASSEMBLY

It is nearly always necessary to redesign a product if it is to be assembled by robots, since existing assemblies make use of human dexterity, visual sensing and intelligence to cope with objects that a robot would find hard to handle. In the process it is often possible to improve the design and in particular to reduce the number of parts. This section lists some ways in which difficulties can arise for a robot, together with ways of avoiding them. Note that most of this applies to assembly by hard automation as well as by robots.

Part numbers and types

To reduce the time taken to assemble the product, the total part count should be minimized. For an existing product there may be some completely redundant parts if the design is a modification of an earlier one or if common subassemblies are used for a range of products.

If a range of similar products is made which were originally designed independently they may use components which are not identical but could be. This would reduce the number of types of part which the factory as a whole has to handle. Similarly, within a single product there may be variations among components (e.g. several kinds of screw) which can be eliminated. Replacing these with a single type allows a reduction in the number of kinds of feeder and tooling, and eases stock control.

It may be possible to reduce the part count by replacing several parts by one which combines their functions, such as a screw with a captive washer. Fasteners such as nuts can be eliminated by replacing a threaded item by a clip-in one.

Component design for mechanical handling

Many conventional parts tangle easily or jam inside each other; this can be reduced by redesign (Figure 7.23).

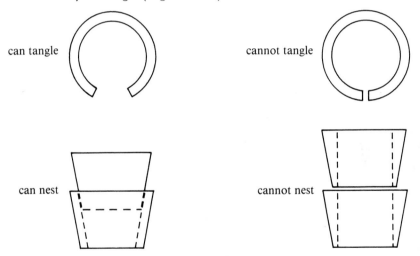

Figure 7.23 *Redesign of components to avoid tangling.*

Problems can arise with the automatic feeding of parts if there is no easy way of telling which way round a nearly symmetrical part should be. There are two cases. In the first only one end or face of the part is functional.

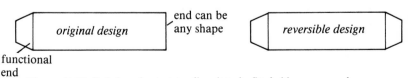

Figure 7.24 *Redesign of a part to allow it to be fitted either way round.*

137

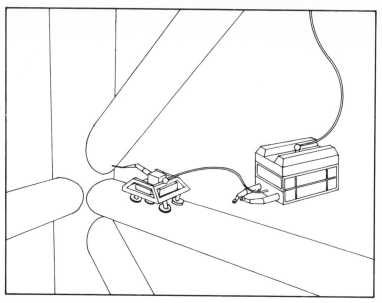

Figure 7.25 *Robotic inspection of welds in an underwater structure. The probe arm is carried to a welded junction by a crawling vehicle, which has been brought to one of the structural members by a remotely operated submersible, seen at right.*

In this case if the other faces are made identical with the functional one it does not matter which way round the part is fitted (Figure 7.24). In the second case a definite orientation must be achieved. The solution here is to add a feature which the feeder or robot can detect and use to orientate the part. It may not be easy to do this, and one of the advantages of intelligent robotics over earlier technologies is that sensors such as vision can be used to determine component orientation.

A third class of problem with parts is that they often deviate from their specification to the point where they will no longer fit. Traditionally, the assembly worker uses his skill and judgement to reject useless parts or to modify his procedure to make use of out-of-tolerance ones, e.g. by applying more force or adding extra washers. This is generally not possible for a robot, so a tighter control of component dimensions may have to be kept.

Finally, parts may have to be redesigned so that a simple gripper can hold them firmly.

Design of the assembly process

The human way of assembling a product should not necessarily be reproduced by a robot, since it is optimized differently. For example, the product should not have to be turned over while it is being assembled, as this needs an active jig. Parts should be added from a single direction if possible. The sequence of assembly should avoid wasted movements such as frequent tool changes.

After assembly the product should be kept in a known orientation for further automatic handling, protected from damage and tracked as it progresses through the factory. If a particular subassembly is used in a range of products it should be committed to a particular product as late as possible; this allows flexibility of scheduling and minimizes the consequences of faults.

New applications for industrial robots

Several trends are apparent in the use of industrial robots, such as

1) the extension of existing methods to new applications,
2) the incorporation of more sensing, especially vision,
3) the increasing use of the methods of AI (item (2) is one example of this),
4) attempts to handle more difficult workpieces, which are flexible or of irregular or varying shape.

These trends seem to be a more or less permanent feature of robotics: at almost any time since the first industrial robot was installed, efforts could have been identified in each of these areas. However, these trends have become stronger in recent years as the supporting technologies such as computer vision have improved.

EXTENSION OF EXISTING METHODS TO NEW APPLICATIONS

There are still unexploited niches for robots of limited sensing ability, which need mainly the development of robots with the right size, geometry and cost. Some of these tasks could be done by existing robots but they are too expensive. An example is education where several companies now make cheap and simple robots; another is the use of robots for laboratory work. Much work in analytical laboratories is repetitive, and indeed fixed automation exists for common instruments: some, such as scintillation counters, have mechanical sample handling devices built in. However, as with manufacturing, there are tasks which are done in batches too small to justify hard automation but large enough to be tedious; there is also the safety issue, since laboratories often deal with poisonous, infective and radioactive substances.

For these reasons robotics is starting to be applied to laboratory work. Some tasks could be done by existing industrial robots, but these are generally too big or expensive. Yet the performance needed is often greater than that of educational robots. Therefore there is interest in developing robots of intermediate quality and cost, and which can be interfaced to electronic balances, pumps, spectrometers and so on. A robot which has been developed for this application, among others, is the Universal Machine

Intelligence RTX, a medium-sized SCARA robot, with six axes driven by DC servomotors, a repeatability of 0.5 mm and a load capacity of 4 kg.

In Britain the Laboratory of the Government Chemist is organizing developments in laboratory robotics.

MORE SENSING

Several recent developments have been mentioned in other chapters. For example, visually guided arc welding has only been developed in the last two or three years, and the commercial availability of vision systems for extracting two-dimensional position and orientation is a post-1980 development, although a few visual inspection systems were available earlier.

Vision and tactile sensing are being applied to new problems almost daily. An example of interest because it departs from the usual fixed terrestrial location for an industrial robot is a project on the use of robots under water on offshore structures. This belongs in this chapter and not in that on teleoperators since the robot will be programmed and will carry out its program automatically.

The aim of this project at University College London is to use a programmed arm for inspecting welded joints in undersea structures, primarily drilling rigs. The joints concerned are where horizontal and diagonal bracing members meet the legs; both braces and legs are steel cylinders, so each weld is defined by the intersection of two cylinders. The principle is to determine the position and orientation of each cylinder by touching it in several places with the robot; having then fitted a mathematical model of each cylinder to the spatial data it is possible to calculate the equation of the arc of intersection of the two cylinders. A testing instrument can then be scanned along the weld by the robot. Figure 7.25 shows the arm, on a crawling platform, standing on one member while probing a joint. The robot is deployed from a remotely operated submersible.

ARTIFICIAL INTELLIGENCE

The applications of AI to robotics are described in Chapter 11.

HANDLING DIFFICULT WORKPIECES

Industrial robots were developed to handle hard objects of fixed shape, and as long as they had simple two-jaw grippers and little or no sensing it was impossible to handle other objects and materials. Yet there are many potential applications of robots for workpieces of the following types:

1) flexible solids (e.g. sponge),
2) flexible sheets (e.g. cloth),
3) flexible wires, ropes, threads or tubes,

4) compressible or elastic materials (generally also flexible),
5) inhomogeneous materials (e.g. meat),
6) live animals,
7) plants (as in fruit picking).

Some tasks involving such objects and materials are as follows:

1) cloth cutting, joining and other handling,
2) handling flexible electrical wires (stripping, soldering, crimping, welding, labelling and tying into looms),
3) laying glass fibre reinforcement,
4) sheep shearing,
5) crop harvesting,
6) loading textile machinery.

There is, of course, a long tradition of automatic handling of some of these. For example, sewing and knitting machines perform complex thread and yarn handling operations, and packaging machinery can handle intractable materials with remarkable facility. Many of these operations which are done routinely appear on first sight to need dexterity and intelligence; other examples are wrapping a block of butter, or putting the barbs in barbed wire.

This suggests that almost anything can be automated given enough ingenuity and effort. The key is to break the operation down into its most fundamental elements and to analyse each using an adequate physical model. Practically all objects, however complex or formless at first sight, can be modelled as an assemblage of cylinders, blocks and so on, each having physical characteristics (density, stiffness, tensile strength etc.) which, if not constant, at least lie within limits. Even the skin of a sheep can be mathematically modelled, although the model is complex and approximate.

It should be added that an alternative to the analytical approach is to observe some empirical fact and to exploit that. Many materials if left to themselves will consistently behave in some way which may seem idiosyncratic but which can be exploited. A material's natural tendency to fold or not fold, or to float or sink, can be used as a basis for handling.

Robots are already well established in some areas of flexible material handling such as assembling wiring looms and wiring up computer backplanes. This was possible mainly because suitable tools already existed for manual use, such as wire-wrapping guns with automatic stripping.

Robotic textile handling is being developed by, for example, Hull University. Special techniques are needed for separating the top layer of a stack of cloth pieces. One method is to blow a jet of air over the cloth. Air penetrates the porous cloth (knitted cotton), forming a bubble under the top layer, which starts to vibrate and partially detach from the layer below. The lower jaw of the gripper is positioned just over the edge of the stack (Figure 7.26); the vibrations eventually make the tip piece flip over the edge of the jaw. The infrared beam between the jaws can discriminate between the presence of zero, one and more than one layer, and the infrared backscatter sensor detects the vibrations of the cloth. These sensors are used to control the grasping cycle; the robot can try again if it fails the first time. In this way the target of 99% successful acquisition of a single layer has been achieved.

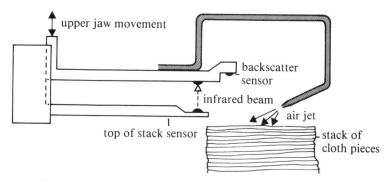

Figure 7.26 *Gripper for picking up an individual piece of cloth from a stack.*

Perhaps the most exotic application of industrial robots to be attempted so far is sheep shearing. In this project of the University of Western Australia a sheep is strapped to a restraining table while a special manipulator moves a clipper over the sheep's skin. The clipper must 'fly' just above the skin while keeping at a fixed angle to the surface; it must make parallel overlapping passes over the sheep's body.

The system uses a combination of modelling and feedback. A mathematical model (the 'software sheep') of the sheep's surface generates the basic pattern of clipping passes. Touch and proximity sensors allow fine adjustment of the tool trajectory so that the edge of the cutter keeps a constant distance from the skin.

The original version of the system, called ORACLE, was first tested in 1979; an improved system called SM is shown in Figure 7.27. The sheep is held in a support and manipulation platform which presents the underside and then each side to the shearing robot. The SM has a telescopic arm hanging from an overhead gantry, and a special 'elephant trunk' wrist incorporating a coupled pair of universal joints to provide the rather special angular motion characteristics required of the cutter.

Contact of the cutter with the sheep is detected by electrical conduction through the skin. Sensing its distance from the skin is a subject of continuing research, as all the methods available such as capacitance sensing and ultrasonics have problems in accurate sensing through a fleece, which has very variable properties.

A second, simpler project in automated shearing is being carried out by Merino Wool Harvesting Pty Ltd of Adelaide. Both these projects are supported by the Australian Wool Corporation.

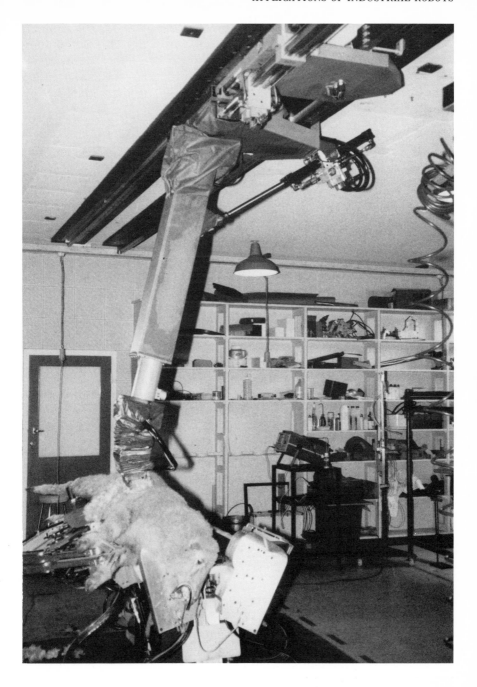

Figure 7.27 *The SM sheep-shearing robot (courtesy of the University of Western Australia).*

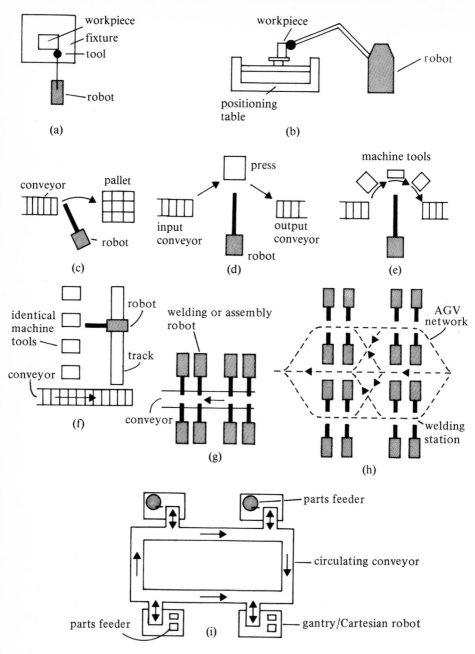

Figure 7.28 *Common workplace configurations: (a), (b) robot doing a long continuous path job such as grinding or arc welding — in (b) it controls a positioning table; (c) palletizing; (d), (e) robot serving one or more machines and conveyors in a fixed sequence; (f) one robot serving several machines in parallel; (g) a conventional line for welding; (h) welding stations, each with four robots, served by an automated guided vehicle fleet; and (i) a manufacturing cell.*

Integration of industrial robots into the workplace

A robot may be almost isolated, or completely embedded in a large machine. Some of the most common arrangements are shown in Figure 7.28; of course, there are endless variations on these. The first six ((a)–(f)) are single-robot installations. In (a) and (b) the robot does a long and complex job and earns its keep regardless of how the workpieces are transported; they may be loaded into the jig or positioning table by hand. Figure 7.28 (c) shows the opposite: the robot's sole function is a single loading task. Figures 7.28 (d)–7.28 (f) represent more complex loading functions in which a robot services several machines; in (d) and (e) it services them in a fixed sequence and in (f) the machine tools are all identical and work in parallel on separate workpieces.

Figure 7.28 (g) is a conventional assembly, spot welding or spraying line. In (h) the workstations are isolated and serviced by AGVs.

Figure 7.28 (i) is a manufacturing cell. The workpieces or asemblies circulate on a closed path. Since they can make repeated passes in any order through the workstations, a closed-path cell is flexible and can use fewer stations than a straight line.

The installations shown all have fixed robots except for (f) where one robot moves on a track, but the robots may well be on tracks or gantries in some of the other examples, either because the workpiece is very large or for extended tracking of a moving workpiece.

The rest of this section discusses some factors which have to be taken into acount when integrating robots into a production process.

TRACKING

If a conveyor moves continuously past a fixed robot then, if it is slow enough and the robot has continuous path control, the robot can track the workpiece over a zone called the tracking window (Figure 7.29). The speed of the belt is v and the length of the window defined by the intersection of the conveyor (or workpiece width) and the working envelope is L_1 on the near side and L_2 on the far side. The robot can dwell on a spot on the workpiece for L_1/v if it is on the near side and L_2/v on the far side. A more realistic problem is to calculate whether there is enough time to do a job such as spot welding.

If the robot cannot do its task in the time determined by the tracking window and the conveyor speed, or if it does not have continuous path control, then tracking must be done by moving the robot on a parallel track. The alternative is a discontinuous conveyor.

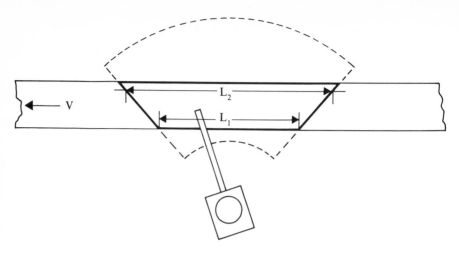

Figure 7.29 *A tracking window.*

WORK CELL CONFIGURATIONS

Figure 7.20 shows a work cell of type (i) of Figure 7.28 in more detail. This is a printed circuit board filling cell. Small pallets each carrying a few boards circulate on a roller conveyor. Although not shown, pallets could enter and leave the loop by conveyors, or they might be brought in a vertical stack by an AGV or manned trolley, from which one at a time would be fed on demand.

The conveyor has two robot workstations each having a siding into which a pallet can be diverted. Once in a siding the printed circuit board is located accurately by a jig while components are inserted by robot. If one station runs out of components or otherwise fails, the pallets can continue to circulate while the other robot goes on working.

Figure 7.30 shows a work cell for assembling cylinder heads. The conveyor forms a closed rectangular path as before, but the assembly jigs are not in sidings. This line involved the design of a special machine for compressing the valve spring while a cotter pin is inserted. It uses eight Pragma 3000 Cartesian robots and assembles 30 different models of cylinder head, which can be mixed on the line.

TRANSPORT OF COMPONENTS AND ASSEMBLIES

Most systems for filling printed circuit boards, whether robotic or using hard automation, take components such as integrated circuits from magazines, and parts with axial leads from bandoliers, loaded by human workers. The transport of these component carriers consumes relatively little time and often is not worth automating. This is true of assembly in general: the assembly is transported by a conveyor system while the small

Figure 7.30 *A flexible assembly line for diesel engine cylinder heads (courtesy of Fairey Automation Ltd).*

components needed at each work station are manually put into the feeders. The pressure towards more complete automation, however, will cause some factories to go to automatic transport of even small parts throughout the factory. This implies a need for a second set of automated machines, possibly robotic, supplying parts to the assembly robots.

Machining presents less of a problem, since at each workstation material is removed rather than parts being added, so apart from the removal of swarf and replacement of tools the only transport needed is that of the workpieces.

FEEDERS FOR ASSEMBLY ROBOTS

As remarked earlier, some parts are fed to the robots in magazines and bandoliers. Note that this implies that the parts have first to be loaded into these carriers: an automation problem for their supplier. Many parts, however, arrive loose in boxes and have to be orientated and presented to the robot by devices such as vibratory bowl feeders. In vibratory feeders vibration causes the parts to move in tiny jumps up a track starting in the bottom of a bin of parts. The track is provided with guides, slots and holes which divert or turn round any parts in the wrong orientation. Parts are also sometimes supplied in individual depressions in a container, like an egg box, which is a variant of the magazine.

CONTROL AND COMMUNICATION

Within a cell all the machines have to be synchronized at least, and flexibility of use may require individual robots and machine tools to be loaded with programs or machining parameters. In the example of Figure 7.31 a programmable logic controller (PLC) synchronizes the conveyor, machine tools and robots of a work cell, while programs are loaded from a central computer over a network. In true computer-integrated manufacture the programs, including that of the PLC, would be derived from the design of the item to be made; this design would originate in a CAD system.

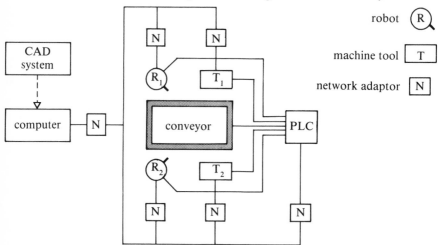

Figure 7.31 *Control connections for a flexible manufacturing cell.*

The network shown uses equipment and cabling designed for this purpose, providing features such as high noise immunity, error checking, a high data rate, long distance operation, many attached devices and compatibility with standard PLC and machine tool interfaces. It is manufactured by Texas Instruments.

In a highly automated factory there may be higher level computers in the hierarchy of control, responsible for scheduling the whole operation, directing jobs to particular cells and so on.

Bibliographic notes

The applications listed on page 117 are described in detail in Engelberger (1980). A useful introduction to assembly is Owen (1985). There are now several books on flexible manufacturing and related subjects, a series of conferences on assembly automation (see for example Heginbotham (1985)), and at least two journals, the *International Journal of Advanced Manufacturing Technology* and the *Journal of Manufacturing Systems*.

Chapter 8
Teleoperated Arms

Introduction

Modern teleoperators originated with the need to handle radioactive material. When it became necessary to separate the material from its handler by a shielding wall, mechanisms had to be developed to transmit an operator's hand movements to a grasping device on the other side of a barrier. The first, all mechanical, telemanipulators for this purpose were developed in the 1940s, notably by a team led by Ray Goertz at Argonne National Laboratory in the United States. Machines of this kind are still in extensive use today and greatly outnumber motorized teleoperators. Radioactive handling continues to be the main application of teleoperators, but other uses have emerged and are described later.

This chapter describes the different types of control commonly used for teleoperated arms, and the mechanical features which teleoperators do not share with industrial robots. It goes on to discuss applications. An extended treatment is given by Vertut and Coiffet (see bibliographic notes).

Methods of control

As the potential for computer-aided control begins to be explored, and as experience with industrial robots is absorbed, it becomes possible to consider new kinds of control schemes, and something will be said of this later. Traditionally, however, the aim of the teleoperator has simply been to give the user sole charge of the movements of the gripper in an ergonomically satisfactory way. This is usually done in one of three ways:

1) mechanical master–slave telemanipulators,
2) powered telemanipulators with open-loop control,
3) bilateral servo manipulators.

MECHANICAL MASTER-SLAVE TELEMANIPULATORS

In these the movements of the operator's hand are reproduced by a linkage of cables, tapes or shafts connecting a slave arm with a gripper to a master arm ending in a hand control. Because the linkage is mechanical its length must be limited and the slave must be in a fixed position relative to the master. This restricts the use of these telemanipulators to the handling of materials in fixed installations. The mechanical connection allows the user to feel the forces on the slave, although with some degradation due to friction, and so there is some degree of feedback. The mechanical advantage of the linkage for each joint need not be unity (although it usually is), so some force or distance multiplication is possible. Figure 8.1 shows a typical mechanical master–slave manipulator. Figure 8.2 shows the principle of the cable transmission for five of the six degrees of freedom of such a manipulator. The sixth is produced by rotating the whole device about the long axis of the sleeve through the wall. Note that since both master and slave move up and down together their weight must be balanced by a counterweight. In practice the counterbalancing is often more complicated than shown here; indeed, there are many other complications — to allow a position offset between master and slave and to permit the slave to be straightened at the shoulder for withdrawal, all while carrying around ten cables or steel tapes through the shoulder joints and connecting tube. The slave arm may be enclosed in a flexible sleeve sealed at the gripper, or, if it uses shafts instead of cables, these can pass through seals.

This geometry, with five revolute joints and one telescopic, is the most common but others are known.

Before leaving this section, it should be noted that a related device is the micromanipulator used to manoeuvre objects under a microscope. The function of the mechanism here is to scale down the user's finger movements to an amplitude suitable for cellular manipulation.

POWERED TELEMANIPULATORS

These use actuators, nearly always electric or hydraulic, to drive the joints of the manipulator. In the open-loop or unilateral case there is no force feedback to the operator. Cranes and excavators are controlled in this way. The operator's control device may take the form of a master arm whose joint angles are measured and drive the corresponding joint servos in the slave arm, or it may be a set of pushbuttons or valve levers, in which case the operator may or may not be able to vary the velocity of a joint continuously. Powered manipulators are used whenever mobility is required, as in a large radioactive cell, or whenever the load is too great for a mechanical telemanipulator, or when a mechanical linkage would be too difficult to arrange.

Servo control of unilateral telemanipulators

Both open-loop and bilateral (described later) manipulators use servo control. This section explains the basic form of servo remote control used in unilateral manipulators by examining a single joint. It should be noted

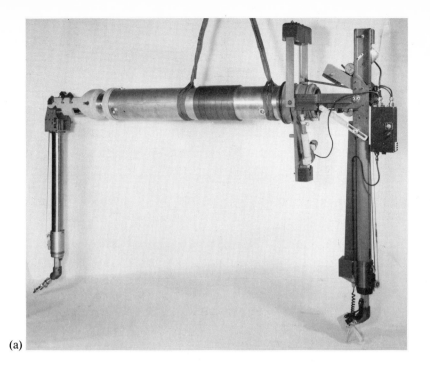

(a)

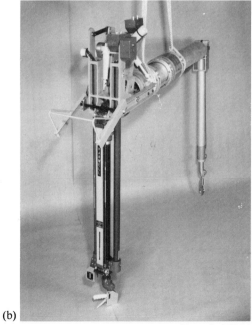

(b)

Figure 8.1 *A mechanical master-slave manipulator, the Harwell GT 3000 Mk 1: (a) general view; (b) view of master end showing chain and cable pulleys in the shoulder (courtesy AERE Harwell).*

153

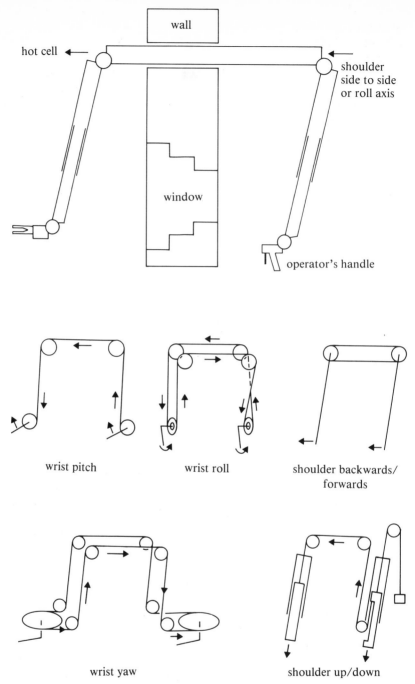

Figure 8.2 *Cable transmissions for five of the joints of a mechanical master-slave telemanipulator. The sixth degree of freedom is obtained by allowing the whole manipulator to rotate about the long axis of the tube through the wall.*

that not all axes must use the same control method. For example, it is sometimes convenient if the master hand control follows the slave motion in the vertical plane but does not move much horizontally, responding rather to force in this direction.

A servo for unilateral control of one joint is shown in Figure 8.3. It is open-loop in that there is no electrical or mechanical feedback from slave to master, but there is a position feedback loop within the slave half of the system. In this loop, if the position signal from the slave potentiometer is equal to that from the master, there is no motor movement.

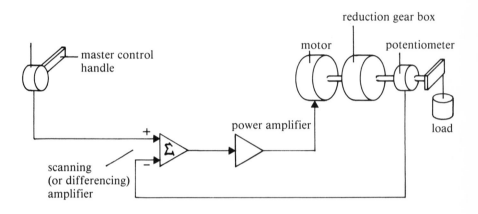

Figure 8.3 *Servo system for an open-loop or unilateral manipulator; the angle of the slave shaft follows that of the master shaft.*

The slave servo loop can be of several kinds. For example, the motor torque may be proportional to the position error signal, or its velocity may be, and for optimum response the velocity of the load should be taken into account as well as its position. The control task is essentially the same as for a joint of a programmed industrial robot. Also, as explained in the section on bilateral manipulators, its details depend on whether the motor can be back-driven or not.

BILATERAL SERVO MANIPULATORS

These are powered but closed-loop in the sense that the force on the slave arm is fed back to the master arm for the operator to experience. This implies that the system is symmetrical in that the master can be moved by a force applied to the slave as well as vice versa, just as with a mechanical master–slave telemanipulator. The key ways in which it differs from the mechanical system are that (a) there can be amplification of force or change of scale between master and slave, allowing a very large or strong or small

155

slave, and (b) since the connection is not mechanical the slave can be distant or mobile.

There are many possible kinds of bilateral control system. The master can be designed to take position or force as its input, the controlled variable at the slave also being position or force, and similarly for the return loop in which slave position or force results in a position or force at the master. This makes 16 possibilities, even without considering other possible control variables such as velocity. In practice only a few of these have been found useful. Some common cases are described shortly. First, however, we must note that, as with unilateral control, there is a great variety of servo systems which can be used in either part of the telemanipulator, whether for force or position. The only distinction dealt with here is between devices such as torque motors which act as a transducer of an input signal such as voltage to torque or force, and devices such as some motor–gearbox combinations which are in effect rigid and can be thought of as transducers of input signal to position. The second kind cannot be back-driven and so, if an applied torque must be able to rotate the output shaft, feedback is needed to sense this torque and to drive the motor in the appropriate direction.

The different sorts of servo which result from making this choice of actuator type are shown in Figure 8.4. Both a position servo and a force servo are shown for each case. Each of the four servos can be used for either the master or the slave, and although the master and slave often use the same type they are not bound to.

A postion servo ((a) and (b)) maintains the output shaft angle equal to the input commanded angle regardless of load. A torque or force servo ((c) and (d)) maintains the output torque or force proportional to that commanded regardless of position. It assumes that some torque, such as that provided by the hand of the operator, opposes the torque produced by the motor, so as to stop the shaft moving. When this is true, the signal from the feedback torque transducer balances the input signal, so there is no error signal and the shaft does not turn.

The second summing amplifier in (a) and (c) is needed because the motor is back-drivable and so, even when there is no error, current must still be supplied to the motor to keep up a torque, whereas in (b) and (d) friction in the gearbox supports the load when the servo is stationary, so the motor current can be zero as long as there is no error signal. (In these examples there is a sustained load. If this is not the case, e.g. if the load is purely inertial, then even with a back-drivable actuator there is no need to maintain a steady torque unless there is a position error.)

Three different examples of bilateral manipulator control systems will now be presented. In each case there is a master servo and a slave servo, each chosen from one of the four types just discussed.

In the first, both servos are of type (b): the result is called a position-position bilateral telemanipulator (Figure 8.5). Given back-drivable actuators, type (a) servos would be used.

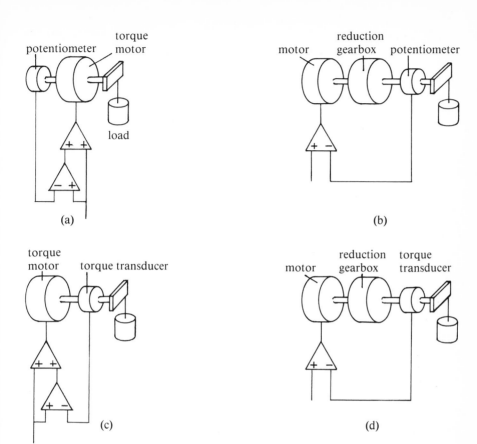

Figure 8.4 *Servo systems for bilateral manipulators. Four types are shown, resulting from the choice of actuator types and whether torque or position is controlled. A bilateral manipulator uses two such servo systems for each axis, one for the master and one for the slave: (a) position servo, back drivable motor; (b) position servo, non-back drivable motor and gearbox; (c) torque servo with back drivable motor; (d) torque servo, with non-back drivable motor and gearbox.*

When the user moves the master arm through a certain angle the slave follows, just as with a mechanical master–slave manipulator. Because the two sides are in effect rigidly coupled, the operator experiences the forces on the slave even though they are not explicitly fed back.

The next example is called a position–force system. The slave is controlled as before, but the signal fed back to the master is torque (or force for a prismatic joint). This is shown in Figure 8.6. If all the components were perfect this would give the same result as the position–position servo, for the slave arm must still follow the master and the master still experiences the torque or force on the slave, but in practice, faced with a variety of situations, some designers have preferred position–position and others position–force.

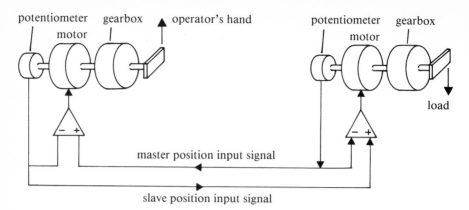

Figure 8.5 *Servo system of position-position bilateral telemanipulator.*

Note that with these manipulators, because they are symmetrical or effectively so, the direction of movement is determined by the net force or torque, i.e. the difference between the force or torque applied to the master and that applied to the slave, and so the slave will drive the master if the force on it is greater.

The third type of bilateral manipulator is rather different in that the slave position does not copy the master position but rather is controlled indirectly; the operator applies a torque or force to the input, which is translated into torque or force (and therefore acceleration if the load is free to move) or velocity of the slave. This has the advantage that the master arm need not move, and so does not need a large volume to move around in. It need not be an arm at all and can take the form of a pistol grip with up to six sensing axes (three forces and three torques); sometimes a pair of handles each with three degrees of freedom is preferred. A possible system is shown in Figure 8.7, using type (d) servos.

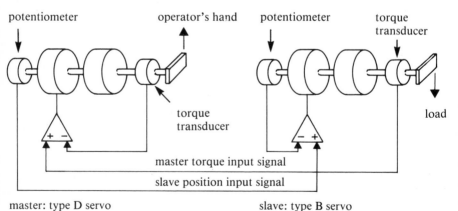

Figure 8.6 *A position-force bilateral telemanipulator.*

Special characteristics of teleoperators

This section draws attention to some differences and similarities between teleoperators and industrial robots and discusses various aspects of teleoperator organization. To begin with, we may note that an arm designed as an industrial robot can be teleoperated if suitable controls are provided; it will then fall into the category of open-loop powered telemanipulators.

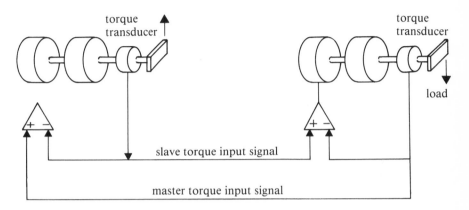

Figure 8.7 *Servo system of a torque-torque (or force-force) bilateral telemanipulator.*

DESIGN CRITERIA FOR TELEOPERATORS

Arms intended as telemanipulators tend to have different design criteria from industrial robots. Since human vision and sometimes touch can be used to position the gripper, the linkage may allow a lot of elastic deflection and some backlash, unlike an industrial robot which usually lacks external sensors and must have the precision to position its gripper without feedback. The deflection of a mechanical master–slave manipulator under full load may be a few centimetres. The acceptability of such deflections allows the use of cable and steel tape transmissions, which are rarely found in industrial robots. Cables and tapes are used so much because they provide a relatively simple way of driving the several joints of the arm, wrist and gripper while also allowing withdrawal of the slave arm through a narrow tube. Cable drives are often used even when a telemanipulator is powered, so the actuators can all be mounted at the shoulder, with cable drives to the wrist, to reduce the mass and bulk at the end of the arm.

Some aspects of the performance of a teleoperator are assessed differently from those of an industrial robot. Speed is assessed not absolutely but in terms of how much slower an operation is when compared with the same operation done directly by human hand. This time ratio can be as little as 1.5 or as high as 100 depending on the complexity of the task and the type of feedback available.

Some performance considerations apply to both teleoperators and industrial robots but have been more clearly appreciated for teleoperators because in their case they are more critical or more obvious. One of these is the volumetric efficiency. The slave arm will itself occupy a volume which cannot therefore be used, and the ratio of this volume to that which can be accessed by the manipulator is a measure of its inefficiency. This is of particular concern if a limited space is available, as in one of the 'hot cells' or 'caves' for handling radioactive material. A related issue is that of manipulator coverage: it is important to minimize the space in a cell which cannot be accessed by any manipulator.

VEHICLES AND TRANSPORTERS

Telemanipulators are often mobile whereas industrial robots are usually fixed, although the same concepts of mobility apply to both. It is often necessary to move a telemanipulator from place to place, or to increase its working volume by mounting it on a platform which slides on rails; this is sometimes true of an industrial robot as well. The extra movement can be regarded as just another axis of the manipulator, but it is convenient to divide the teleoperator system into three levels (which may not all be present in a particular machine): the vehicle, the intermediate transporter and the telemanipulator. The vehicle and telemanipulator are self-explanatory; the intermediate transporter is usually a Cartesian arrangement of prismatic joints. In nuclear installations it often has three axes and is built into a reactor hall or a storage building, so there is no need for a vehicle. An example is shown in Figure 8.8. The intermediate transporter may carry two telemanipulators instead of one.

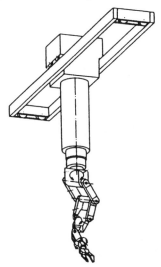

Figure 8.8 *A telemanipulator mounted on a three-axis transporter.*

Applications of teleoperators

REMOTE HANDLING OF RADIOACTIVE MATERIALS

Much of the development of teleoperators has been stimulated by the need to handle radioactive materials, and mechanical master–slave manipulators are used for few other purposes.

The nuclear application imposes several constraints on the design of manipulators. The most obvious are due to radiation itself. If the operator simply has to be protected by a barrier from direct radiation from a fixed source then the manipulator linkage can be fitted over the wall and can be as large and as complex as is convenient. Usually, however, manipulation has to be done in a zone which becomes contaminated with radioactive substances and so must be sealed off by a barrier which not only absorbs radiation but forms an airtight container. In this case any mechanical linkages must pass through airtight seals. Further, it is desirable for the slave arm to be capable of withdrawal through its access hole. If this is not possible, provision must be made for remotely disconnecting it within the cell and removing it by some other route. A manipulator for access through narrow holes is shown in Figure 8.9.

As well as constraining the mechanical design in this way, radiation damages many materials and components, such as hydraulic oil, flexible seals, electronic circuits and motor winding insulation, so these must be eliminated, shielded or made easy to replace. This accounts for the continued predominance of mechanical manipulators.

A further problem is that visibility is poor. A fixed manipulator can be viewed through a window in the cell wall; these are very thick and absorb and scatter a lot of light, giving a rather dim, distant and restricted view of the work area. The alternative (a necessity if manipulation is done far from the cell wall) is to use television cameras, which are sensitive to radiation and offer limited resolution and a depthless image.

REMOTE HANDLING OF EXPLOSIVE AND TOXIC MATERIALS

In Britain one of the best-known examples is the use by the Army of remotely controlled vehicles for bomb disposal. These tracked vehicles are usually fitted with a simple manipulator having just two or three joints. Since the steering of the vehicle is used to position the end effector in azimuth, it does not make sense to adhere to a rigid division between vehicle and telemanipulator.

Industrial robots are sometimes used in munitions work, as much for safety as for economic reasons. This can be considered as a case of remote handling which is not teleoperation in the usual sense since the robot is programmed and not under continuous human control.

There is likely to be an increasing use of teleoperators for dealing with dangerous materials such as asbestos.

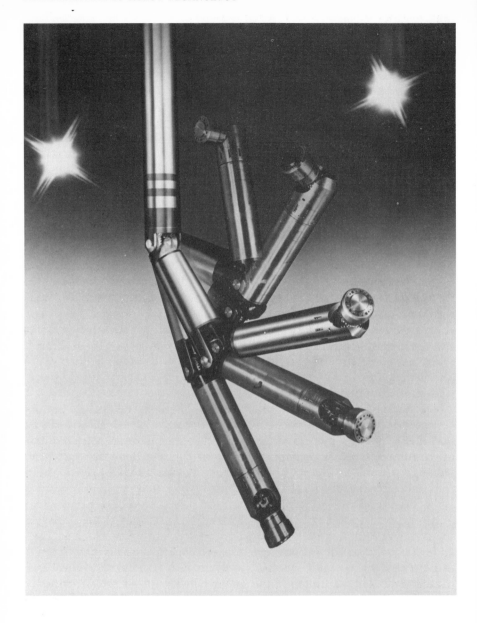

Figure 8.9 *A powered manipulator, which can be programmed or teleoperated, for reaching through narrow holes such as a reactor fuel element channel (courtesy of Taylor Hitec Ltd).*

TELEMANIPULATION OF HEAVY OBJECTS

The forging of steel billets has always required mechanical assistance, originally in the form of large tongs supported by a hoist. More recently the need to handle heavy and sometimes hot items such as engine blocks has led to the development of telemanipulators such as the Andromat, designed by General Electric and made under licence by Automation und Steurungstechnik GmbH of Kassel, and the Lamberton Robotics Scobotman series. In both of these the operator sits in a cabin alongside and rotating with the master arm (Figure 8.10). These manipulators provide force feedback. The slave replicates the master arm's position in the vertical plane, but the side to side or arm azimuth input responds to force rather than displacement. The control handle on the end of the master arm has three torque inputs controlling the rotational joints of the wrist. The slave arm normally ends in a heavy duty pincer-like gripper, but tools such as grinders and cutting torches can be fitted. The largest machines can carry a load of several tonnes. They use hydraulic actuation.

As well as general purpose manipulators like these, there are machines dedicated to particular tasks, e.g. replacing facing plates in a grinder. These grinders have the form of a drum about 10 m in diameter and the facing plates weigh tens of kilograms. The manipulator is inserted through a hole on the axis of the drum and transfers facing plates to the periphery.

UNDERWATER TELEOPERATION

Submersibles and bottom-crawling vehicles, both manned and unmanned, are used for a variety of tasks such as recovering torpedoes and crash debris, collecting mineral samples, the salvage of sunken ships and the construction and maintenance of marine structures. The subject of robot submersibles is covered in Chapter 11, but their telemanipulation aspect is dealt with here.

There are two cases: manned and unmanned. In the former the main reason for using a telemanipulator is that it is desirable to keep the crew compartment at atmospheric pressure, and although it is possible to fit a pair of rigid jointed diving-suit arms to the vehicle, so that the vessel becomes a cross between a submarine and a diving suit, these arms are bulky, of limited manoeuvrability and difficult to seal without excessive friction. Instead, telemanipulators are usually fitted — these are often hydraulic (see Figure 8.11) as there is no problem of radiation resistance and it is relatively easy to make a hydraulic system resistant to high pressure seawater. The operator views the slave arms through windows.

Unmanned submersibles have the advantages that there is no danger to a human occupant and that, because there is no crew compartment with its life support and safety requirements, the vehicle can be smaller and cheaper. The cost of this is the difficulty of providing adequate feedback, which must be by television. There are two problems: television produces even poorer images under water than in the air, and its high bandwidth imposes great demands on the link to the human operator who is usually on a surface ship. Either a cable capable of handling television signals must be provided or a wireless, usually acoustic, transmission method must be used, in which case the bandwidth is drastically limited and only low resolution or low repetition rate images can be sent.

TELEOPERATION IN SPACE AND PLANETARY EXPLORATION

The first example of telemanipulation in space is the arm on the American space shuttle (see Figure 8.12(a)). It is distinguished by its great length (15 m) and its unusual environment in which objects have mass but no weight, so inertial forces are all-important. Two other dynamic peculiarities are that since it is so long and lightly built it cannot be assumed to be rigid for control purposes, and that since it is mounted not on the solid Earth but on a free-floating platform the effect of its reaction against its mounting is significant. The arm is controlled by a pair of three-axis joysticks with force feedback.

The operation of manipulators on planetary surfaces presents a unique problem: the time delay in the control loop if they are controlled from Earth makes the usual control methods difficult or impossible. The problem is compounded by the low bandwidth of interplanetary radio links, so that television frames can be sent only infrequently. It becomes necessary either to work very slowly or to introduce some degree of automatic (programmed) control. The first method is acceptable for simple operations on the Moon, where the round-trip delay is only 2.5 s, but hopeless for Mars where it is tens of minutes.

The tasks performed by teleoperation so far have been driving a vehicle (Lunokhod) on the Moon, and digging up samples of the surface material (Surveyor on the Moon and Viking on Mars). Because of the problems just mentioned, the Viking sampling arm was controlled by directing it to a site of interest and then starting a program which automatically dug up a sample and transferred it to the analysis station — see Figure 8.12(b).

TELEMANIPULATORS FOR THE DISABLED

One of the most laudable goals of robotics is to restore a measure of function to those whose limbs are deformed, have been amputated or are useless because of paralysis. Prosthetic (limb replacment) and orthotic (limb assistance) aids have, of course, a long history, but in recent years an element of telemanipulation has been introduced. They are worn by the user who can to some extent use bodily movement to compensate for their deficiencies. For severely paralysed people it may be better to mount a complete telemanipulator on a table, wheelchair or bed, or even on a mobile robot, and to provide the user with special controls operated by head, mouth or

Figure 8.10 *Industrial heavy load manipulators: the Lamberton Robotics Scobotman (courtesy of Lamberton Robotics Ltd).*

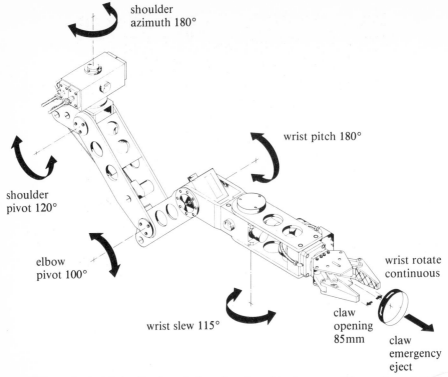

shoulder
azimuth 180°

wrist pitch 180°

shoulder
pivot 120°

elbow
pivot 100°

wrist rotate
continuous

wrist slew 115°

claw
opening
85mm

claw
emergency
eject

Figure 8.11 *A hydraulic telemanipulator for submersibles (courtesy of Norson Power Ltd).*

eye movements, or voice (Figure 8.13). Research has been carried out at various establishments on several aspects of the problem, despite chronic difficulty in obtaining funds and scepticism on the part of the intended users. (One problem is that people want aids for use in the home to be like domestic appliances in being neat, cheap, clean, easy to use and needing virtually no maintenance, whereas manipulators are generally designed for industrial uses where all these requirements are commonly sacrificed for good performance.) The three most sustained reasearch efforts have been those of Paeslack and Roesler in Germany, the French Spartacus project and a series of projects supported by the United States Veterans' Administration. Figure 8.14 shows an industrial robot being used as an aid by a handicapped person.

Computer-assisted teleoperation

There are several ways in which computers can enhance the capabilities of a teleoperator. First, note that, although a computer can implement a digital servo system, in this case it just invisibly replaces an analogue servo; this does not count as computer assistance.

A straightforward example of computer assistance is the ability to call

(a)

(b)

Figure 8.12 *Space and planetary teleoperators: (a) the space shuttle arm holding a small satellite; (b) the Viking soil sampling arm on Mars (courtesy of NASA).*

167

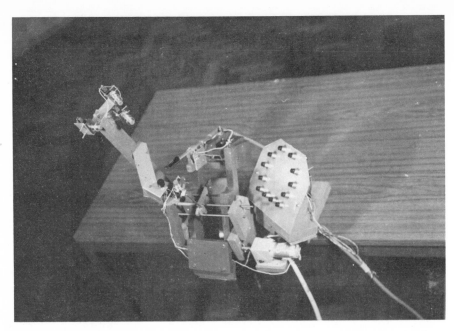

Figure 8.13 *Input devices for a paralysed user of a manipulator. On the right is a group of 12 pushbuttons which can be pressed with a mouth stick; in the middle is a mouth-operated joystick which can be moved out of the way when not in use; and at left is a group of head-operated switches.*

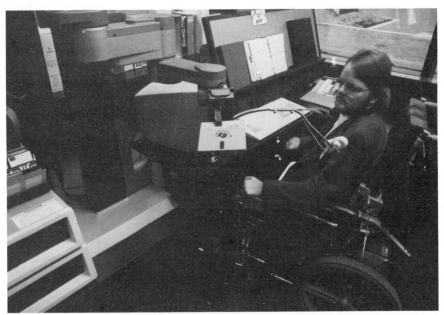

Figure 8.14 *An industrial robot controlled as a computer-assisted telemanipulator by a paralysed person (courtesy of Universal Machine Intelligence Ltd).*

up programmed sequences of actions, as with the Viking sampler or some of the aids for the paralysed. These programmed actions may include branches conditional upon sensor signals, or may take a sensor reading as an input. An example is the Stanford University manipulator for the disabled, whose hand can 'fly' parallel with a surface, automatically rising to cross obstacles, by sensing the surface using optical proximity sensors. In another aid, developed at Queen Mary College, London, the process of grasping was guided by feedback from touch and force sensors in the gripper and a light beam between its jaws to detect the presence of an object. The use of such sensors does not need a computer, but without one little flexibility of behaviour is possible.

In the example just described the computer provides a means of constraining the actions of the arm according to sensor signals. A related function is the generation of constraints by software. This means that the computer calculates whether an undesirable condition is about to arise, such as degeneracy of axes (see Chapter 2), or a collision, or excessive force. To the operator it feels as if the arm has encountered an invisible barrier.

The use of a computer also makes it possible for the user to work in a coordinate system ungoverned by the geometry of the arm (up to a point). This is useful if the gripper must be moved in a straight line or rotated about a point (as in turning a handle), actions requiring the coordinated motion of several joints. This is not a problem, at least in theory, with mechanical or bilateral powered telemanipulators as the slave follows whatever the master does, but can be a problem for unilateral powered manipulators if the controls do not readily allow the velocity control of several joints at once.

The ability to command a movement such as a straight line in a useful coordinate system such as a Cartesian one, regardless of the geometrical configuration of the robot, and have the robot automatically generate the necessary joint motions is called resolved motion control. If it is the velocity in some desired coordinate system which is commanded, the term 'resolved motion rate control' is used.

An indirect way in which computers can assist teleoperation is in enhancing or supplementing television images of the working area. There are several possibilities, such as aids to depth perception or showing a computer graphics simulation of objects which the cameras cannot see well. This would need the system to store an accurate model of the work area and to be able to generate a perspective view of it from a given direction.

Such a 'world model' could also be used by the computer for planning collision-free arm trajectories and vehicle routes. Research on these subjects is in its infancy.

Bibliographic notes

For an extended treatment of teleoperators see Vertut and Coiffet (1985). The future of robotics (mainly teleoperated) in the nuclear industry is discussed in Larcombe and Halsall (1984). Thring (1983) also deals extensively with teleoperation.

Chapter 9

Mobile Robots

Introduction

The machines which may be regarded as mobile robots are very diverse. Looked at in the most general way, a mobile robot is a machine which can move as a whole in a controlled way with some degree of autonomy. As with a manipulator, its possible motions can be analysed in terms of degrees of freedom: it may fly or float in a three-dimensional medium, in which case there are three translational and three rotational degrees of freedom to be controlled (not counting any internal joints); or it may be constrained to follow a surface, in which case there are two translational degrees of freedom and one rotational degree of freedom (yaw or azimuth); or it may be confined to a track or pipe, when there is one translational degree of freedom and there may or may not be one rotational degree of freedom (roll). By the use of internal joints a surface robot may be able to control its body attitude in three dimensions.

This chapter is concerned with robots confined to surfaces or moving through a medium. It begins with land vehicles. One category not discussed here is the AGV. Because these are well established in industrial use and have a specialized technology of their own, they are dealt with separately in Chapter 10.

Land surface robots

The means of locomotion of most importance for robots are wheels, tracks, legs and air cushions. The last of these will not be discussed, as so far there has been little interest in robot hovercraft, although this may well change once military robotics gets into its stride. Most mobile robots so far, like land vehicles in general, have used wheels since this is the simplest method of locomotion; wheels are abandoned only when there is a pressing reason. Tracks are intended as a sort of portable railway, a temporary smooth road for a wheeled vehicle, which spreads the load on soft ground and bridges gaps. The purposes of legs are discussed later.

ARRANGEMENTS OF WHEELS AND TRACKS

There are three fundamental ways of steering a wheeled or tracked vehicle by ground reaction, and they all need at least two wheels. In the least restricted way (Figure 9.1(a)) all the wheels or tracks can be turned together or differentially, resulting in translational or rotational motion or a combination. Such a vehicle can rotate about its own centre and, being very manoeuvrable, is much favoured for robots in buildings. Three wheels are often used.

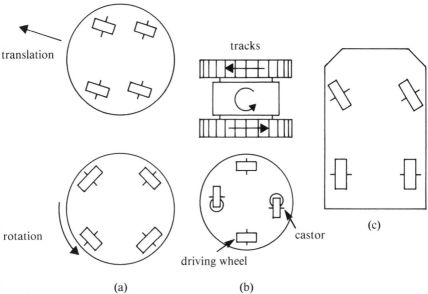

Figure 9.1 *Steering methods for wheeled and tracked vehicles: (a) all wheels steered, collectively and differentially, (b) differential steering, (c) steering at one end only.*

The second way, used by caterpillar tractors and some AGVs, is differential steering (Figure 9.1 (b)) of a pair of fixed wheels or tracks. Again it allows rotation on the spot but the vehicle cannot move sideways.

The third way (Figure 9.1 (c)) is to pivot one wheel or group of wheels while others remain fixed; this is the method used by cars and bicycles. It has rather poor manoeuvrability but simplifies the construction of fast and efficient vehicles. This arrangement is rarely found in robots as turned round in a confined space or passing through a narrow gap tends to need complex back-and-forth manoeuvring.

Unusual wheel and track arrangements

The machines shown in Figure 9.2 and 9.3 use wheels for efficient travel on smooth ground augmented by special mechanisms to allow stair climbing. The triangular wheel-clusters of Figure 9.2 have been used for hand-carts as well as for mobile robots. The three wheels of each cluster are driven by a common motor through a chain or gears. A mobile robot using this principle, designed at the University of Tokyo and called TO-ROVER, is now produced by Toshiba under the name AMOOTY.

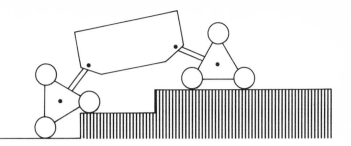

Figure 9.2 *Wheel-clusters for stair climbing.*

The machine shown in Figure 9.3 is a remotely controlled robot designed by Hitachi for maintenance in nuclear power stations. It has five vertically telescoping legs each having a steerable, powered wheel on the end. On a smooth surface it simply rolls along on the wheels. To climb stairs it successively extends and retracts the legs to match the stair profile while rolling along one stair-tread at a time.

Variable geometry tracked vehicles have been designed in a variety of forms intended to confer good mobility on rough ground. The design of Figure 9.4 allows a vehicle to adopt a wide range of postures. That of Figure 9.5, which is similar to the design of a Sno-Cat except that each of the four tracks can be individually raised, lowered and steered, was envisaged at the Jet Propulsion Laboratory, Pasadena, for a Mars rover.

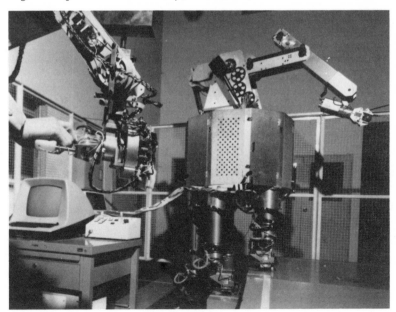

Figure 9.3 *A stair-climbing robot with a wheel on the end of each telescopic leg (courtesy of Hitachi Ltd).*

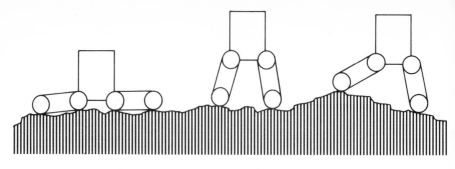

Figure 9.4 *Variable geometry tracked vehicle.*

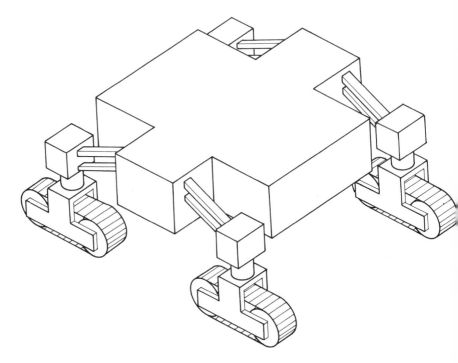

Figure 9.5 *A vehicle with four independently suspended and steered tracks.*

NAVIGATION FOR LAND VEHICLES

Navigation is used here to mean determining the position of a mobile robot. It is sometimes defined to include route planning and following, which are here treated separately. There are many possible methods, which are now listed.

Teleoperation

If the operator can see the vehicle directly or see its surroundings through an on-board television camera, the robot need have no autonomous navigation ability. It may happen, however, that vision is obscured, so supplementary methods are advisable.

Dead reckoning

Odometry (travelled-distance measurement) by counting wheel revolutions together with direction finding using a magnetometer or gyrocompass, or differential wheel rotation, is often used for wheeled indoor robots. It always shows a steadily increasing error due to wheel slip and tyre elasticity, and therefore the vehicle needs to call at intervals at known locations at which the navigation system can be reset. Dead reckoning is of little use on rough ground except as a coarse or emergency method.

Inertial navigation

An inertial navigation system uses a kind of dead reckoning in that position in three dimensions is calculated as the double integral of acceleration, measured by accelerometers on a gyrostabilized platform. Inertial navigation systems of quality are extremely expensive and have found little favour for land vehicles.

Tracking from a fixed base; beacons

Triangulation using radar, optical or acoustic methods can be used to locate the vehicle relative to fixed stations and the information can be transmitted to the vehicle. A related technique is for the radar or other sensor to be on the vehicle and to detect beacons which may be lights, reflectors or transponders.

A special case of the beacon method is so-called hyperbolic navigation systems such as Decca and LORAN which use radio phase comparison methods. They need the working area to be within the coverage of the chain of radio beacons.

Satellite navigation

The equipment for a satellite navigation system, which uses radio transponders to measure the distance to satellites of accurately known position, is small enough to fit to some of the larger mobile robots.

Map matching

If a robot has sensors such as radar, an optical rangefinder or stereo vision it can scan the surrounding area, producing an image (whether in range or brightness). If this image can be analysed to give a representation of the local topography, this can be compared with a map. The image analysis stage is difficult, particularly since prominent features in the image such as large nearby trees may not be shown on the map, and the map may use symbolic representations of bridges and so on which cannot be compared directly with the image. (The attraction of this system is diminished unless it can use, in digital form, standard maps such as Ordnance Survey maps.) The problem is made easier if the robot keeps a continuous record of where it is and can use dead reckoning to estimate what sort of scene to expect. This subject is in its infancy.

Wall following

This is a special case of map matching. A robot inhabiting a world of rooms and corridors can carry a map showing just the walls, and can match this with its sense impressions of the walls in its neighbourhood gained by rangefinding or by odometry while following a wall.

ROUTE PLANNING

Assuming that the problem of navigation can be solved, a mobile robot, except when teleoperated, must work out a route to its current destination. Route planning can be regarded as a branch of AI, and so is discussed in Chapter 11.

CONTROL AND COMMUNICATION

Mobile robots generally have a hierarchical control structure in which on-board software handles the lowest levels such as interrogating sensors and driving motors, and often the intermediate levels such as following a path or even planning a route. The highest level of control is often remote control by a human operator. In this case the on-board software must be able to receive commands by radio or some other link and to carry them out, maintaining a safe course of action even when communication is interrupted.

Communication is usually by radio, infrared or cable. Each has its problems. A cable can become entangled, limits range and imposes drag, but is cheap and simple and is sometimes useful as a way of pulling the robot out of a dangerous place if its own mobility fails. If radio is used it may need bandwidth for television signals, and so it has to be ultrahigh frequency or microwave and there may be legal restrictions on its use and problems of obstruction and interference. An antenna pointing mechanism may be necessary. Infrared is strictly a line-of-sight method and so is even more prone to obstruction than radio, but is immune to most forms of interference. A combination of methods is needed for the highest reliability.

As implied above, mobile robots generally have some degree of autonomy, even if limited. This subject is discussed in Chapter 11.

SENSORS FOR MOBILE ROBOTS

A mobile robot may need to measure several quantities unimportant to a manipulator, such as body orientation, velocity and acceleration, and the distance to objects.

Body orientation and angular rates

The terms 'attitude' and 'orientation' are often interchangeable, but it is sometimes helpful to distinguish yaw or azimuthal angle from pitch and roll. On a level surface yaw is distinguished because it is a navigational quantity rather than reflecting an internal disposition. Here the term attitude is used to mean pitch and roll, and azimuth or compass orientation is used for angle about the vertical or surface-normal axis. Orientation by itself is taken to include all three axes.

There are various ways of expressing orientation, e.g. Euler angles. Choosing a consistent system of reference is important if the changes in angle are large, but it is often adequate to use terms such as pitch and roll rather loosely. As used here pitch is defined as the angle of rotation about a transverse axis fixed in the body, and roll is the angle about a longitudinal axis fixed in the body.

The usual sensors of pitch and roll are gyroscopes or damped pendulums. Pendulums can be relatively cheap but are sensitive to acceleration. They can be instrumented to give a continuous measure of angle, or to give just a binary indication of departure from the vertical by a certain amount.

A better instrument (a typical accuracy is $\pm 1°$) is the vertical gyro, which provides both pitch and roll signals from a single instrument, usually as potentiometer outputs.

Angular rate in pitch and roll may in principle by obtained by differentiation if the angle is known with enough accuracy, but if the rates are high (tens of degrees a second) or if high accuracy is needed it is better to use a rate gyro. A separate gyro is needed for each axis.

For azimuth, gyroscopic sensors are available but expensive. An alternative is the fluxgate magnetometer. This is commercially available or can be built without too much difficulty. An accuracy of about $1°$ is possible at mid-latitudes away from large masses of iron. Another method is to sense orientation relative to a beacon or set of beacons, which can also give other information such as position. The beacons may be optical, acoustic or radio depending on the range and propagation conditions. A beacon system is obviously only useful when it is possible to place beacons in advance — it has been used for the underwater navigation of submersibles, for instance.

Body position, speed and acceleration

These quantities have a global and a local interpretation. On a global scale body position means map position and its determination is navigation, an issue not specific to walking machines. Speed, for navigational purposes, is speed averaged over some fairly long period, and acceleration is not directly relevant.

On a local scale these quantities are aspects of the problem of achieving smooth and stable locomotion. The most direct way of sensing body position is by finding the range to known objects. A second way, for legged robots only, is to calculate the coordinates of a body reference point from the measured joint angles and the positions of the supporting feet. It is likely that the foot positions will not be known in an absolute sense, but at least this enables the position of the body relative to the local surface to be determined.

Velocity and acceleration can again be calculated from the wheel or joint velocities and accelerations, but it is often preferable to sense them directly. Velocity can be sensed by Doppler radar or sonar aimed at the ground ahead, or, in principle, from analysis of television pictures, although this is not yet an established technique. Acceleration in the range relevant to dynamic control can be sensed by accelerometers.

Terrain scanning

A mobile robot, whether legged or not, needs to scan the ground ahead to avoid obstacles and, if it has legs, to select footholds. Terrain scanning can perhaps be dispensed with on smooth ground, but this is probably a rare condition for legged robots. It is also unnecessary if it has a human driver who is prepared to control foot placment but, as with riding an animal, it is better if the driver can concentrate on other things while the machine chooses its own footholds. If a walking vehicle is an autonomous robot then terrain scanning becomes connected with navigation, as the machine must select not just a locally suitable path but an effective route to its destination.

Because of the range and resolution requirements, terrain scanning is usually done at visible or near-infrared frequencies rather than by sonar or radar. There are several possible methods, including

1) monocular scene analysis (using a single television camera),
2) binocular scene analysis,
3) triangulation ranging using a television camera or other detector to locate the spot of light where a laser beam meets the ground,
4) time-of-flight ranging,
5) interferometric (phase comparison) ranging.

These methods all have limitations. In particular, the scene analysis methods, although the subject of much research, are far from being mature enough for reliable use on a walking machine.

Perhaps the most sophisticated system yet to be designed for a walking machine is the continuous wave phase comparison system for the OSU

Adaptive Suspension Vehicle. This principle of range measurement is used in surveying equipment. A beam, in this case of infrared light from a gallium arsenide laser, is amplitude modulated and phase comparison of the transmitted and reflected beams gives the range. The maximum range is about 10 m (although by using multiple modulation frequencies the technique can be extended to a range of kilometres) and the resolution a few centimetres. The beam is scanned in a raster pattern covering the ground ahead of the vehicle.

In Britain the Royal Armament Research and Development Establishment (RARDE) has developed a scanning time-of-flight rangefinder for the 100 m range region. It uses a pulsed solid state laser and a polygonal mirror scanner to cover a field of 45° in azimuth and 30° in elevation, and generates a 55 × 56 pixel range image.

TYPES AND APPLICATIONS OF MOBILE ROBOTS

The correlation between the applications and the physical forms of mobile robots is not perfect, but for most applications only one or two physical types are common. This section lists the main applications of robot surface vehicles and in each case describes the types of robot used.

Education and Research

Mobile robots of the turtle type (Figure 9.6), connected as peripherals to microcomputers, are used to enliven the teaching of programming, of computer principles, and of elementary geometry and other aspects of mathematics. Turtles usually have two driving wheels driven by stepper motors, with differential steering, and so dead reckoning can be used. Some turtles are equipped with photocells, touch sensors, bar code readers and so on, and place little constraint on the intelligence of the programs which can be written.

The next step up is a self-contained vehicle large enough to carry batteries, more sensors and an on-board computer. These vehicles are often about the size and shape of a dustbin, convenient for manoeuvring among obstacles, and approximating the ground area of a human being. Differential steering with two driving wheels is sometimes still found, as are stepper motors, but other wheel configurations, such as three powered and steerable wheels, also occur. The drive motors are often DC servomotors. These robots are used for research in subjects such as navigation for AGVs (see Chapter 10) and AI.

The most famous AI research robot was Stanford Research Institute's Shakey (Figure 9.7). This project was ambitious for its time (1966–1972), especially as there were no small computers and so Shakey had to be controlled by a fixed computer (a PDP-10 in the final version) over a radio link.

The project's goals were to develop techniques and concepts in AI to enable a robot to function autonomously in a realistic environment. The

Figure 9.6 *The BBC Buggy, a 'turtle'-type robot; it is connected to a microcomputer by a ribbon cable (Courtesy of Economatics Ltd).*

main thrust of the research was in the area of planning actions, solving problems, recovering from mistakes and learning in the sense of storing and re-using plans. Although the project was not directly concerned with vision, the robot used a television camera as one of its main sensors, so some computer vision software had to be provided and integrated with the rest of the system.

Shakey was driven by two large stepper motors. As well as the two driving wheels there were two load-bearing castors. The main sensors were the camera (with motorized pan, tilt, focus and aperture control), mechanical touch sensors, driving wheel shaft encoders and a rangefinder. The robot rolled about in an environment consisting of several rooms connected by doors and containing a few large, regular objects such as cubes and wedges.

Shakey was eventually able to solve problems such as 'push the three boxes together'. Such a problem, the system's highest level goal, would be typed in using a subset of English. The robot then generated a plan using a problem solving program called STRIPS. Given a description of an initial state and a goal state, expressed in predicate calculus (as was Shakey's world model), STRIPS would work out a plan (by a kind of

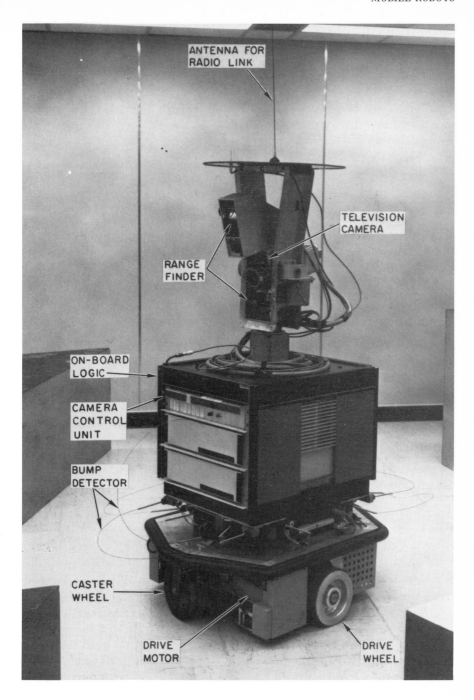

Figure 9.7 *Stanford Research Institute's Shakey (courtesy of SRI International).*

181

theorem proving) which consisted of a chain of the standard operations which could be done by the robot. These were called intermediate level actions (ILAs); an example is 'go to a specified place'. An ILA was made up of several low level actions in a suitable control structure.

Once a plan, a chain of ILAs, was generated it was executed, one ILA at a time, under the supervision of an executive program called PLANEX, which monitored execution and could adjust a plan or abandon it if a problem arose. Plans could be generalized to some extent and stored for subsequent re-use on similar problems.

The project was discontinued in 1972. Research on intelligent mobile robots continued at a low level of activity compared with that on manipulators. Among the projects of the 1970s were Berkeley's JASON, which was used for experiments in problem solving, typically involving pushing boxes through doorways, and a series of robots at Queen Mary College, London. The emphasis here was on models of learning. One of these robots is shown in Figure 9.8.

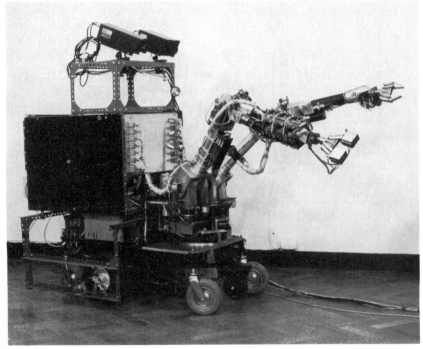

Figure 9.8 *The Queen Mary College Mark 5 mobile robot (courtesy of C.M. Witkowski).*

Simple mobile robots were built, and still are, as student projects in many colleges. More recently, robots have been built to carry on where Shakey left off. A notable example, at least in engineering terms, is the Carnegie–Mellon University rover; another is the HILARE robot of the Université Paul Sabatier, Toulouse. This has more advanced visual processing, and more modern (knowledge based) methods of world modelling, planning and problem solving. Its world model includes a network representation of the sort shown in Figure 11.2, together with a more geometric representation of individual rooms. It can build up the model by exploring its environment. Like other modern robots it carries some on-board computing power but still communicates with a fixed computer for much of its processing.

Remote Handling

Mobile robots for remote handling of radioactive, explosive and other dangerous materials are generally teleoperated and carry manipulators. An example is shown in Figure 9.9.

Unlike turtles and research robots they must be able to carry loads and to exert forces and so need high stability; also, they may have to cross obstacles. Therefore they usually use four or more wheels or else tracks. Even so, there is a danger of their becoming top-heavy and falling over if the manipulators hold heavy loads too far out.

Remote handling robots have been built with internal combustion engines for long-endurance outdoor use, but most are electric, powered by batteries if free ranging or supplied through the control cable if tethered. In future legged robots are likely to find a use in remote handling. An example of such a use is working in a reactor building contaminated by an accident; it may be necessary to climb stairs and step over obstacles, which cannot be done satisfactorily by a wheeled or tracked vehicle. Figure 9.3 shows a hybrid vehicle for this purpose, with wheels on the end of legs. Another application is bomb disposal in buildings. Figure 9.10 shows a mock-up of a stair-climbing robot.

A variable geometry robot intended for remote handling among other uses is shown in Figure 9.11.

Military Mobile Robots

Apart from bomb disposal, which can be regarded as an example of remote handling, several uses for mobile robots have been proposed:

1) mine clearance,
2) mine laying,
3) transport of ammunition and other supplies to battlefield units,
4) as decoys,
5) testing for chemical and biological agents and radioactivity,
6) electronic countermeasures,
7) reconnaissance,
8) sentry duty,
9) as unmanned tanks and other weapons carriers.

183

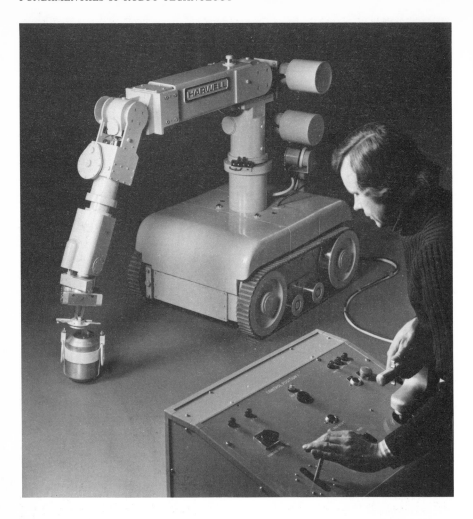

Figure 9.9 *The Harwell ROMAN tracked teleoperated robot (courtesy of AERE Harwell).*

These require the robot to be largely autonomous, i.e. teleoperated for little of the time if at all. Therefore it must be able to determine its location and to plan and follow routes. Such robots will at times have to make use of roads and tracks shared with other users, as well as being able to travel across country. This will need advanced imaging sensors, probably in addition to more global navigational aids. For many military applications the robot will have to be able to recognize targets, threats and friendly units.

The on-board intelligence needed to interpret the signal from imaging sensors in this kind of environment is very great, and the more advanced functions will need a lot of research.

Work on unmanned military vehicles is under way in several countries.

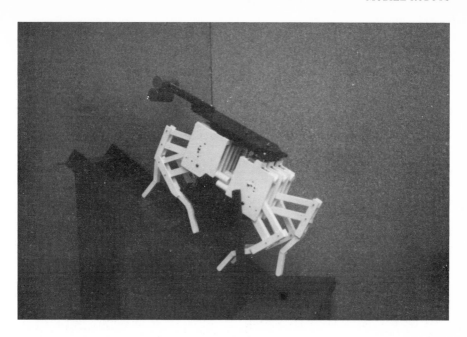

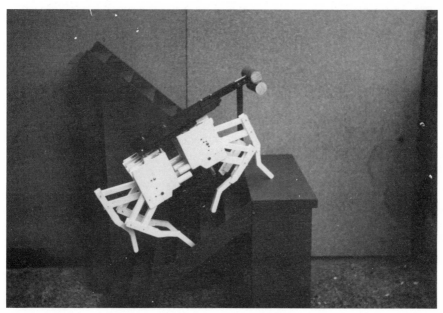

Figure 9.10 *A 1/5 scale model of a stair-climbing robot being developed for industrial load handling, bomb disposal and as a walking chair. Each pantograph leg pivots about a vertical axis. Balancing on stairs makes use of a counterweight on a flexible boom; also, the payload platform can slide relative to the chassis.*

The United States Defense Advanced Research Projects Agency's Strategic Computing Program includes a substantial effort in this area. Some subjects within this are mine clearance, ammunition transport, patrol/sentry robots and support for the walking robot research of American universities.

In Britain the RARDE has a research programme called mobile autonomous intelligent device (MAID). Research is regarded as falling into four areas: mobility, sensing, effectors such as grippers, and intelligence, which includes knowledge representation, mission planning, navigation and route finding among obstacles.

Military mobile robots are currently being designed around multi-wheeled (six being common) chassis, or with tracks. Figure 9.12 shows how extra wide tracks can be used to produce a 'bellyless' vehicle whose advantage is that it has almost no body surface to get stuck on obstacles. RARDE has built a prototype vehicle of this kind.

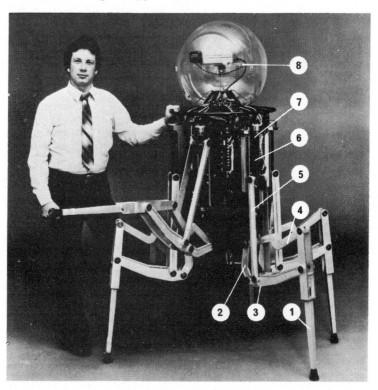

Figure 9.11 *The Odetics Inc Odex I six-legged robot:* **1** *leg created from a parallelogram* **1** *to* **2** *and* **3** *to* **4;** **5** *rod for control of upper leg (3 to 4);* **6** *screw and nut for the control of* **5;** **7** *multiturn potentiometers for the servocontrol of the linear actuator* **6;** **8** *television camera.*

A different class of military robot is the walking truck, designed for ground too rough even for tracked vehicles. Figure 9.13 shows Ohio State University's six-legged Adaptive Suspension Vehicle.

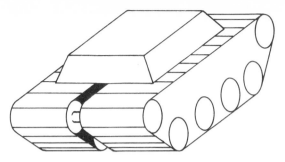

Figure 9.12 *The 'Bellyless' wide-tracked vehicle.*

Figure 9.13 *The Ohio State University Adaptive Suspension Vehicle (courtesy Ohio State University).*

Fire-fighting and Rescue

Robots are potentially attractive for inspecting dangerous areas in fires and for providing a fire-proof vehicle which can be sent through a fire to rescue

187

people trapped by flames or smoke. An alternative is for the robot to carry supplies such as escape devices, breathing equipment or a portable refuge to trapped people. Such a robot would probably be teleoperated by radio, or, if the building was designed with this in mind, could be a wire-guided AGV. Radio is unreliable in areas with a lot of absorbing material such as reinforced concrete, while wires are likely to be damaged by the fire. A third possibility is wall-following navigation; this would need a map of the building to be prepared in advance.

A robot could also fight fires, carrying extinguishers or the end of a hose; in this case it could be teleoperated by cable. Its range would be limited by its ability to drag the hose, but in some fires a few tens of metres would be valuable.

Construction

An eight-legged robot called ReCUS, made by Komatsu, has been used for surveying the sea-bed in preparation for bridge construction, and has been proposed for other tasks such as levelling mounds of rubble. On land, legged construction machinery could offer access to difficult sites and step over obstacles.

Mining

Mobile robots, usually teleoperated, have been considered for many years, but so far have not been found to be economic.

Planetary Exploration

The first, and so far only, unmanned vehicle to have been used was the Lunokhod landed on the Moon by the USSR. It had eight wire-mesh wheels and was solar powered. It was teleoperated from Earth. Since then several vehicles have been proposed, mainly for Mars.

Legged robots

This section has a different emphasis from those on other mobile robots: it concentrates on the technology of locomotion itself, which could to a large extent be taken for granted in the case of wheeled or tracked robots. This is because getting a machine to walk or run satisfactorily is far from a solved problem and so, although a legged robot is ultimately a system whose designer must deal with navigation, planning and so on, the main focus of interest is on the mechanics and control of locomotion with legs.

Because legged robots are in such an early stage of their development, their usefulness is largely unproved, but there is quite a long list of potential applications awaiting their perfection. Most of the mobile robot applications listed earlier could be satisfied by legged robots, and there are one or two others such as walking aids where legs are necessary by definition. This section describes the current state of development of legged robots.

COMPARISON OF LEGS AND WHEELS

Legs have several advantages over wheels and tracks: they can step over obstacles; they can climb stairs smoothly; a smooth ride can be achieved on rough ground; a leg does not, like a wheel, waste power by always having to climb out of a rut of its own making, and is less susceptible to digging deeper and deeper until the vehicle stops; and feet may do less damage to the ground than tracks or wheels. Figure 9.14 shows how a wheel on soft ground continually experiences resistance owing to its compaction of the soil, whereas after the initial sinkage (which does waste some energy) there is little resistance to the forward movement of a leg. On hard ground there is no difference. It can also be seen from this figure that, whereas soil compaction reduces the tractive force of a wheel, it actually serves to anchor a leg and may increase the tractive force it can generate compared with walking on hard ground.

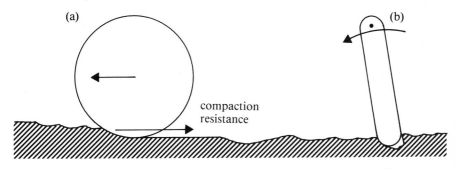

Figure 9.14 *Comparison of wheel and leg: (a) wheel continuously crushing soil; (b) leg experiences little soil compaction after initial sinkage.*

However, legs have several disadvantages. The oscillating motion of a leg wastes power. To achieve the required motion of the vehicle some actuators may have to work in ways which do not contribute to propulsion but waste some power. A legged vehicle is much more complicated than a wheeled one, and much slower.

LEG NUMBER AND ARRANGEMENT

There are several aspects of leg arrangement, of which only a few can be picked out here.

Leg number

High numbers are suitable for heavily loaded, slowly moving vehicles, whereas bipeds and quadrupeds can be fast and agile. Six is a popular number since it allows two alternating tripods and thus a stable but fairly fast gait.

Leg disposition

The legs can be almost vertically under the body as in a mammal, which is fast and efficient, or project from the sides like a reptile, which is low and stable, or can be long and folded like some insects, which allows great agility at the price of reduced strength (Figure 9.15).

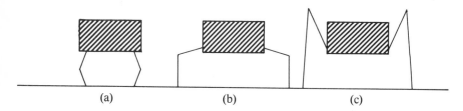

Figure 9.15 *Possible leg dispositions: (a) mammal; (b) reptile; (c) insect.*

Relative leg length

The length and shape of the legs can be varied, perhaps to avoid inter-leg collisions, as in Figure 9.16.

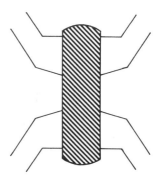

Figure 9.16 *Overlapping legs; the middle legs are longer than the end ones so they can make their stride without mutual interference.*

LEG CONSTRUCTION

First note that a machine which can balance on one leg like a human needs a fairly large foot of controllable orientation; this implies a complex leg with typically six degrees of freedom. However, a many-legged robot which always stands on three or more legs can get away with simpler legs, usually of three degrees of freedom, having a simple foot, if any.

The legs of bipedal robots generally follow the geometry of the human leg, with one or two joints at the ankle, one at the knee and two or three at the hip.

For robots with many legs (multipods), it is desirable for the propulsion and lift motions to be independent; this simplifies control and improves energetic efficiency. It is commonly done using a pantograph to transfer the motion of a pair of actuators, one with a horizontal motion and the other vertical, to the foot (Figure 9.17). The pantograph shown contributes two degrees of freedom; a third can be obtained either by attaching one of the sliding joints to a third slide at right angles or by pivoting the whole pantograph about a horizontal or vertical axis. This method, with a horizontal axis, can be seen in Figure 9.13, and with a vertical axis in Figure 9.10.

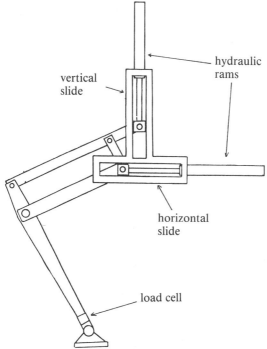

Figure 9.17 *A hydraulically actuated pantograph leg.*

CONTROL

A legged vehicle, like a manipulator, can be regarded as a series of rigid masses connected by joints driven by actuators. The aim of control is to find a set of actuator forces resulting in sustained stable motion of the vehicle. The most basic problem is stability. A machine is said to be statically stable if its centre of mass is above the base of support defined by the feet which are on the ground at the time. A machine can walk slowly using only stable states, but at higher speeds, particularly with few legs, stable states may alternate with brief periods of flight or falling. There may even be no statically stable states at all, although the locomotion cycle as a whole is stable. The control system must ensure that one of these kinds of stability is maintained.

The earliest walking machines were controlled either by a human driver who controlled the legs with master–slave servo systems (Figure 9.18) or by cams and gears. It proved too demanding for a human driver, while the mechanical systems could not adapt to varying ground conditions.

Figure 9.18 *The General Electric Walking Truck. It has been fitted with outriggers to prevent it falling over too far (courtesy US Army Tank Automotive Center).*

The modern era in legged robots began when it became possible to equip a vehicle with an on-board computer. The computer implements a hierarchical control structure whose main levels are as follows:

1) gait generation (deciding which leg to move next, and where to),
2) foot trajectory planning,
3) joint servo control.

The implementation for slow straight-line walking is fairly straightforward. A gait is a repetitive pattern of foot movements. It is often expressed as a gait diagram (Figure 9.19). The most effective gaits for static stability on smooth ground are *wave gaits*. Figure 9.20 shows the gait diagram for a hexapod gait in which waves of stepping sweep from back to front, a gait common in insects. A simple analysis, given the vehicle speed and the stride length, leads to the time of touch-down and take-off of each foot; these times can be used for controlling the raising and lowering of the legs.

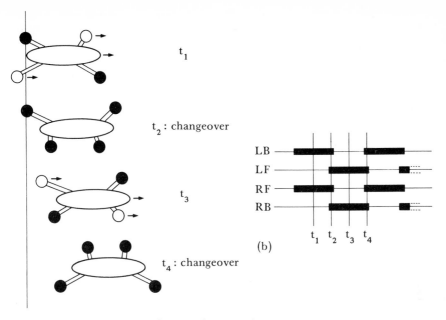

(b)

(a) closed circles represent a foot on the ground

Figure 9.19 *Interpretation of a gait diagram.*

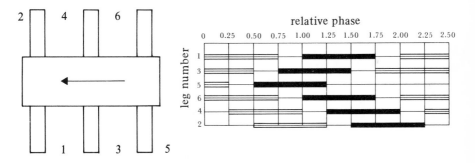

Figure 9.20 *Gait diagram of a hexapod wake gait.*

Gait generation can be extended to turning and to walking on rough ground. Turning requires both a variation among the supporting legs of the length of the stride and a sideways component of foot movement. This can be achieved by computation and servo control or by a mechanical steering mechanism. Walking on ground with obstacles needs a gait in which the next foot to be moved and where it is to be placed are determined by a goal such as maximizing stability. Such gaits are called free gaits. They assume that safe areas for stepping can be identified, which is difficult to do automatically (needing scanning of the ground ahead) and a burden for a human operator.

For dynamically stable machines which walk quickly, run or hop the emphasis of the control problem is on the dynamics of the system of masses, joints and actuators. The locomotion of anthropomorphic bipeds falls into this category, and because of its relevance to human locomotion has been much studied. Bipeds have been built, notably at Tokyo University, which rely on dynamic stability; an example is shown in Figure 9.21. The same group of researchers have extended their work to dynamic gaits in quadrupeds, which resemble a pair of the bipeds of this illustration joined together. At Waseda University a series of bipeds have been built, starting with the WABOT in 1972; the latest machine, WL-1OR, has 12 degrees of freedom (Figure 9.22). It can walk forwards, backwards, sideways and can turn, taking 4.8 s for a stride of 45 cm. Some joints are powered by hydraulic rotary actuators, others by DC servomotors. It weighs 80 kg.

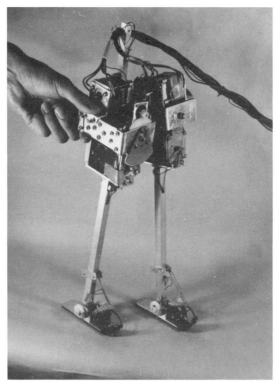

Figure 9.21 *Tokyo University's BIPER 3 (courtesy of Professor H. Miura).*

At Carnegie–Mellon University a series of one-legged hopping machines have been built (Figure 9.23), followed by a quadruped which can run. The hopping machine is of interest from the control point of view because it proved possible to control a highly unstable object quite simply (in principle). The key was the decomposition of the control problem into separate servo systems for height, attitude and forward velocity.

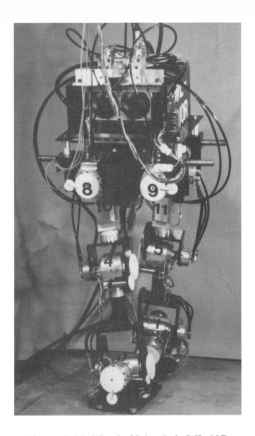

Figure 9.22 *Waseda University's WL-10R.*

CLIMBING ROBOTS

Robots for climbing vertical or overhanging structures have magnetic feet or suckers or mechanical grasping devices. Machines for climbing a smooth surface have been made in France, Japan and Britain. The principle is shown in Figure 9.24. The machine holds on alternately with its body suckers and with its feet, which can slide on longitudinal rails. They are placed against the surface by swinging arms or telescopic joints. The Japanese robot has eight electromagnetic feet, four at the corners of the body and four on a rectangular frame which slides on rails and can rotate about a central axis for steering.

An alternative climbing method is to use a wheeled vehicle with permanent magnets in its wheels or, if under water, suction pumps to hold it against the wall. Such a robot has been built by Mitsubishi for cleaning power station cooling water ducts of marine growths.

Robots have also been designed for crawling through pipes, pressing aginst the walls for adhesion. Propulsion is either by wheels (or tracks) or by alternately extending and retracting body segments like an earthworm.

Figure 9.23 *The Carnegie-Mellon University one-legged hopping robot in action (courtesy of M.H. Raibert).*

Robot submersibles

Robot *submarines,* capable of extended periods of autonomous underwater operation, have been proposed for naval purposes, but *submersibles* which submerge for a few hours at a time are well developed for commercial as well as military purposes. (The term submersible is sometimes reserved for manned craft; here it is used for unmanned vessels as well.) Submersibles are usually tethered to the support ship from which they are operated. The cable provides three functions.

1) Launching and retrieval: the cable supports the submersible during launch and recovery from a ship by gantry or crane.

2) Power supply: submersibles use tens or hundreds of kilowatts, which would need massive batteries if there were no cable.

3) Data transmission: in particular, video signals from underwater cameras need so much bandwidth that wireless transmission is not practicable, although low resolution, low frame rate television pictures can possibly be sent over an acoustic link if data compression techniques are used. Research on this subject is being done by Heriot–Watt University, Edinburgh.

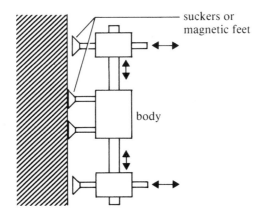

Figure 9.24 *A robot that will climb up a smooth surface.*

There is a range of underwater devices with several intermediate steps between diving bells and suits and unmanned robots. The first step towards the robotic is the atmospheric diving suit or system (ADS). It maintains a pressure of 1 bar (1 atm) so no decompression is needed after a dive. It does not have legs but thrusters (ducted propellers) like a submersible. The arms are jointed pressure-armoured sleeves for the diver's arms.

The next step is to fit telemanipulators instead of diving suit arms. The result is a manned submersible of the kind shown in Figure 9.25.

The next step towards robotization is to remove the diver. Submersibles such as the Mantis, Duplus and Hawk made by OSEL Offshore Systems Engineering Ltd. can be operated either manned or unmanned. Others, such as the Dragonfly and the Rigworker (Figure 9.26), and the UFO of Figure 9.27 do not have a crew compartment and can be operated only as remotely operated vehicles (ROVs).

Manned operation allows work to be done faster and more effectively, mainly because the diver/pilot has much better visual and other feedback, even with a thick window, than a remote operator. However, an ROV is smaller, cheaper and more readily expendable, which is important because of the danger of underwater work. The most common danger is entanglement of the submersible's cable or protuding parts with an underwater structure.

The main features of a tethered submersible are visible in Figure 9.25. The normal construction is a tubular metal framework topped with buoyancy elements. To the frame are attached sealed electronics pods, cameras, lights, thrusters and instruments. If manned the form is dominated by a horizontal cylindrical crew compartment with a hemispherical window dome at one end. The manipulators when fitted are attached low down at the front.

Figure 9.25 *A manned submersible which can also be teleoperated (courtesy of OSEL Offshore Systems Engineering Ltd).*

Both position, in three dimensions, and attitude (roll, pitch and yaw) are controlled by thrusters powered by electric or hydraulic motors. There may be from 4 to 12 thrusters. In some designs such as the UFO (Figure 9.27) they are buried deep in ducts for safety.

The vehicle does not hang by the cable (except during launch and recovery), and the cable should exert as little force as possible. It can be made neutrally buoyant, but in currents will exert a drag which may be the greatest force acting on the submersible.

Tethered submersibles are generally lowered over the side or stern of a ship by a launcher incorporating a boom or A-frame, which swings overboard, and a winch for the cable. When operating in strong currents the submersible is sometimes tethered to a vertical wire so that the thrusters do not have to compensate for all the drag on the vehicle (Figure 9.28(a)). Some submersibles are deployed from a 'garage' (Figures 9.28(b) and 9.29)

supported by a strong heavy cable. The submersible swims out a relatively short distance from the garage, to which it is tethered by a short and
flexible cable which generates little drag.

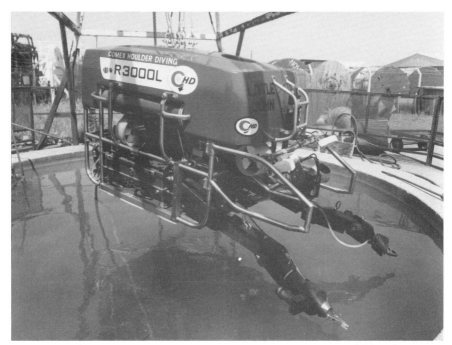

Figure 9.26 *An unmanned submersible robot or ROV (courtesy OSEL Offshore Systems Engineering Ltd).*

For navigation ROVs are generally equipped with a depth gauge, a
gyrocompass or magnetic compass and an echo sounder. The signals from
these are sometimes used by servo systems to maintain a specified depth,
heading or height above the sea-bed. There may also be pitch and roll
servoing using a gyro for the pitch and roll measurements.

These instruments, together with television, are enough for many purposes. If absolute position is important it can be obtained by placing acoustic
beacons (transponders) at known locations on the sea-bed or a structure.
In this case the vehicle is fitted with a computer to calculate position by
triangulation using the range measurements to three or more transponders.

Figure 9.30 shows a typical ROV console used on a ship or drilling rig,
with joysticks for operating the manipulators and driving the vehicle.

The manipulators are sometimes powered by seawater hydraulics.

USES OF SUBMERSIBLE ROBOTS

Some ROVs are used only for observation; such originally was the UFO,
although it can be fitted with a simple manipulator. They can be relatively

Figure 9.27 *A small ROV for inspection, the UFO (courtesy OSEL Offshore Systems Engineering Ltd).*

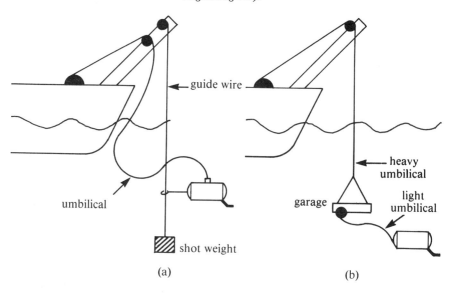

Figure 9.28 *Methods of submersible deployment: (a) use of a guide wire in strong currents; (b) deployment from a 'garage'.*

small and simple, equipped with a range of cameras. Others are equipped from the outset for heavy work; an example is the Duplus, which has two low pressure (5 bar (80 psi, 5 bar)) water-hydraulic manipulators and a third grappling arm. Such a vehicle can weigh more than 1 tonne in air. Duplus can operate at 700 m manned and 1700 m unmanned. ROVs have been designed for special purposes: a famous example is the American CURV for retrieving practice torpedoes and similar objects. It was used for recovering the nuclear bombs lost in the sea off Palomares, Spain, when a B52 crashed.

A list of tasks done by commercial submersibles follows. Some would be done by a manned vehicle in preference to an ROV, but at great depths and in dangerous places there is pressure to eliminate men from all tasks:

1) non-destructive testing,
2) photography,
3) bottom surveying,
4) wreckage location,
5) sampling,
6) cleaning,
7) welding,
8) cutting,
9) placing explosives,
10) operating valves,
11) fitting and removing equipment.

Robots in air and space

Guided missiles of all kinds have many of the characteristics of a robot but on the whole constitute a separate subject. This is more a matter of convenience (since engineering has to be split into fields somehow) than of deep principle. Robots are to some degree felt to be intelligent and anthropomorphic, and a device whose sole purpose is to rush to its own destruction seems a poor embodiment of these qualities.

Another class of machine commonly described as robotic is the remotely controlled aircraft (known as the RPV for remotely piloted vehicle). These range from tiny aeroplanes and helicopters for reconnaissance to the full-size commercial transports occasionally fitted with remote controls for crash testing.

RPVs are by definition teleoperated, and this is currently the normal way of operating a robot aircraft. Teleoperation assumes good and continuous comunication, which is not always easy to guarantee; it also, of course, occupies a pilot all the time it is flying. Therefore there are good reasons for using autonomous or largely autonomous aircraft for purposes such as extended reconnaissance flights.

An example of a recently developed RPV is Lockheed's Aquila. This

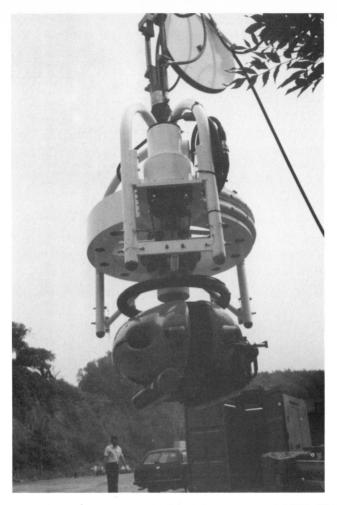

Figure 9.29 *The 'garage' from which the UFO is deployed (courtesy of OSEL Offshore Systems Engineering Ltd).*

2-m long, 4-m wingspan aircraft is driven, as are many RPVs, by a pusher propeller caged to avoid damage on landing. (It is launched from rails and caught in a recovery net.) The Aquila, which has a flight duration of 3 h, can designate targets with a laser for laser-guided weapons or can be used for reconnaissance and damage assessment. It can fly autonomously to a target area and then be controlled from the ground.

Another type of RPV under investigation is a naval patrol airship which could stay on station, autonomously controlled, for long periods.

SPACE

Not all spacecraft must be regarded as robots. A typical communications

Figure 9.30 *The control console for remote operation of the UFO (courtesy of OSEL Offshore Systems Engineering Ltd).*

satellite, for example, contains several servo systems for attitude control, antenna pointing and so on, but it has no behavioural flexibility nor is it in any way anthropomorphic. However, for an interplanetary probe on-board intelligence and the physical ability to take a range of actions can be a great help to survival and performance, and spacecraft with these characteristics can be said to be robots.

(Robot vehicles and manipulators for planetary surfaces are discussed under the headings of teleoperators (Chapter 8) and land surface robots.)

Bibliographic notes

Vertut and Coiffet (1985) and Larcombe and Halsall (1984) contain a considerable amount of material on mobile robots, mostly from the point of view of teleoperation. There are as yet no good books on the technology of mobile robots in general. SRI International of Stanford, California has published a number of reports on Shakey and related subjects.

An introductory text on legged robots is Todd (1985); for a fascinating detailed account of how a particular walking robot was designed and built see Sutherland (1983).

Chapter 10

Automated Guided Vehicles

Mobile robots in general are dealt with in Chapter 9. An important class of mobile robot is the AGV. It is distinguished by being in widespread industrial use, and has a well-defined, although developing, technology, to which this chapter is devoted.

An AGV is a driverless wheeled truck which automatically follows a route defined by, usually, a buried wire or a painted line. In future, AGVs may be able to leave the wire for a short detour round an obstacle (such as another AGV on the same route), and eventually more flexible guidance systems may allow their operation without a wire at all.

AGVs can replace manned vehicles such as fork-lift trucks and tractors, or continuous transporters such as roller conveyors. Manned vehicles have the advantage over AGVs for small volume and irregular operation. Fixed conveyors are best when the path taken never changes and there is a high and steady flow of goods. Some, such as overhead chain conveyors, are the only practical way of doing certain jobs such as dipping objects into liquid baths. AGVs do not compete with these as much as with floor-mounted conveyors.

The main advantages of an AGV system over fixed conveyors are that it dispenses with the space consuming, obstructive and expensive network of conveyors, and is flexible in that laying a new route is relatively simple. Its advantages over manned vehicles are the reduction in labour costs and its ready integration into an automatic goods-handling arrangement or even an unmanned factory.

AGVs also offer the possibility of new ways of organizing production, e.g. by carrying assemblies from one workstation to another, or carrying industrial robots from one workstation to another.

AGVs are currently available in several physical forms:

1) tractors for pulling a train of unpowered trailers,
2) pallet trucks (fork lifts),
3) skid tractors (a low flat-topped vehicle which fits completely under its load, typically. a pallet),
4) assembly platforms (sometimes taking the same form as a skid tractor); a group of these replaces an assembly line conveyor, so the assembly stations do not have to be in a straight line or in any fixed relationship.

An example of a skid tractor is shown in Figure 10.1.

(a)

(b)

Figure 10.1 *An AGV: (a) general view — in this version the four posts on top engage a load; (b) with the cover removed from the wheels at one end and the batteries withdrawn — this version is fitted with a fixture for a particular workpiece, such as an engine (courtesy of Babcock FATA Ltd).*

Automated guided vehicle technology

POWER, STEERING AND GUIDANCE

Most AGVs are powered by lead–acid batteries and DC motors. The powered wheels are also the ones which steer; the remaining, unpowered, wheels may just be castors. Some wheel arrangements are shown in Figure 10.2.

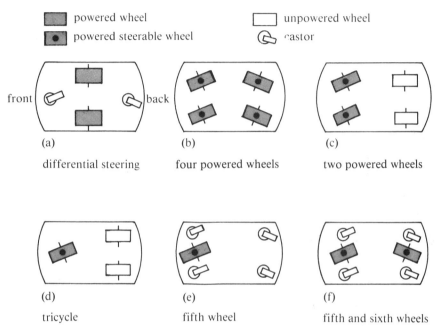

Figure 10.2 *Wheel arrangements for AGVs: (a) differential steering, (b) four powered wheels, (c) two powerer wheels, (d) tricycle, (e) fifth wheel, (f) fifth and sixth wheels.*

The standard guidance method uses a cable grouted into a groove a few centimetres deep cut in the floor. The cable may have one or several conductors. Each conductor carries an alternating signal at a frequency of a few kilohertz or tens of kilohertz and a current of tens or hundreds of milliamperes. The magnetic field produced by the cable induces an alternating voltage in a pair of coils, often in the steered-wheel assembly. The difference between the signals from the two coils is a measure of the lateral position error, i.e. of the distance from the centre of the coil assembly to the buried cable. It is used by the steering servo system to keep the truck following the wire.

This method has been preferred to alternatives such as optical tracking of a painted line because the buried cable is hard to damage and very reliable; also, since it is active it can be switched on and off or modulated with a signal, and it is easy to provide each conductor with a different frequency. By tuning the receiving circuits on the vehicle, one of these can be picked out; thus, at a junction, frequency can be used to distinguish the two branches.

The route must have markers on it at intervals to enable the vehicle to keep track of its position. Such a market is commonly made by diverting the buried cable into a buried coil or loop (Figure 10.3) which induces a signal in a corresponding coil on the truck. Alternatively, buried permanent magnets can be used, with reed switches or Hall-effect sensors on the vehicle.

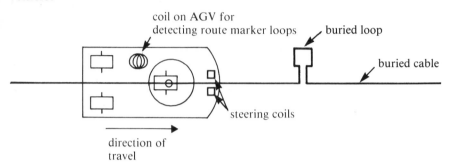

Figure 10.3 *'Through floor' view of the underside of an AGV showing coil for detecting route-marker loops in the buried cable.*

The truck can use the route markers for position finding by simply counting them, or they may be distinguishable in some way, e.g. a pattern of magnets, to give the absolute position within the network. A combination of these approaches is sometimes used in which markers are counted within a branch of the route network, the branch being identified by frequency or other means.

The vehicle may have to signal to the track. For example, in a single-frequency system if the vehicle holds its program on board then, when it comes to a junction, it must switch off the current in one branch of the buried wire so that it will follow the other. Figure 10.4 shows the arrangement of such a junction. When the truck detects the floor magnet it selects branch A or branch B by turning on one of two coils on the truck which operates detectors S1 and S2 in the floor to switch off a short section of either branch B or branch A. Once the truck has passed the junction, which is signalled by S3 or S4, the current can be turned on again. In this system, since only a very short section is ever turned off, many vehicles can use both branches without affecting each other.

In a multifrequency system points are much simpler: the vehicle merely has to select a frequency (Figure 10.5).

ROUTE PROGRAMMING

An AGV has a microcomputer-based on-board control system which carries

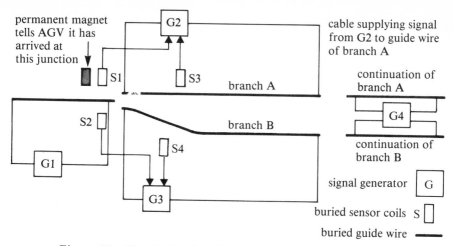

Figure 10.4 *Diverging junctions ('points') in a single frequency system.*

out the servo control of steering and of the velocity and acceleration of the vehicle. It also has to monitor sensors and to drive outputs such as indicator lights. In addition, the control system has to operate the loading equipment if fitted.

The on-board computer is responsible for getting the truck from one specified place to another by executing a route-following program which keeps track of the current location, switches points or selects a route frequency, and stops the vehicle at the required loading stations. If the vehicle continually travels the same circular route the program need not be changed but just restarted when the vehicle reaches the end, which will happen at some fixed reference point on the route. Such a system needs no centralized off-vehicle control.

Each truck can be programmed individually using an on-board keyboard or a portable terminal, or a cassette with an already written program. The program entered is expressed in alphanumeric codes for route, loading stations and so on; the on-board computer translates these codes into the detailed route-following program.

Alternatively, the AGV can be reprogrammed with route and destination details at the start of every journey. (If it happened to keep repeating a single fixed route, the new program would be identical with the old one.) Clearly this requires an automatic method of program loading: the program is transmitted by an inductive or infrared link from a central computer to each vehicle's on-board computer when the vehicle reaches a reference station.

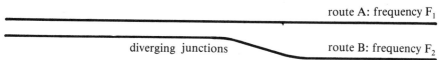

Figure 10.5 *Diverging junction in a multifrequency system.*

ROUTE PLANNING

Route planning means deciding that a journey must be made between two specified places and generating the instructions which must be given to an AGV to make it carry out this journey (and to load and unload itself at the start and destination). For a network like that of Figure 10.6 in which there are many workstations in parallel, it is continually necessary to decide which workstation needs servicing next, to allocate an AGV and to plan a journey for it. Route planning should take into account the need to make efficient use of the AGV fleet, to avoid queues and to cope with breakdowns by, for example, using alternative routes. It may allow options such as the calling up of an AGV when requested by a workstation operator.

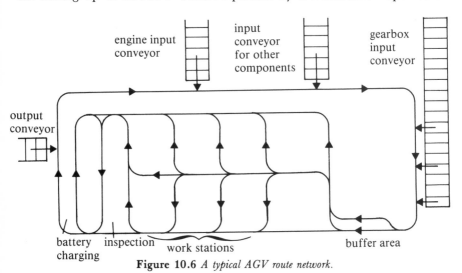

Figure 10.6 *A typical AGV route network.*

In a centrally controlled system route planning is done by the central computer, or possibly by a human supervisor. Route planning will often have to be integrated with the operation of the whole factory.

LOADING AND UNLOADING

AGVs are sometimes topped with a short section of roller conveyor matching those at the loading stations, or they may have powered forks or powered couplings in the case of a tractor; alternatively they may be passive, with the loading done manually or by the stationary equipment.

On reaching a loading station an AGV must first identify the station and then stop at it with the precision needed by the loading equipment. This may involve mechanical guides, or servo control of speed and steering using optical aids. Apart from matching the loading equipment, it will often have to align its infrared communications receivers and transmitters with their counterparts in the fixed structure.

210

Loading may involve actions by actuators both on the vehicle and on the station; signals must be exchanged to coordinate or interlock these actions. These interlock signals can also govern the departure time so that, particularly in a decentralized system, the timing of the whole network can be synchronized with the operations of the workstations and warehouses serviced.

SAFETY; VEHICLE SEPARATION

A loaded AGV can weigh several tonnes, so a number of precautions are taken to prevent them hitting people, objects or other AGVs. First, the speed is limited, often to about 1 m/s. The AGV can be equipped with flashing lights and turn indicators and perhaps a horn. Ultrasonic sonar can detect obstacles up to a few metres away; in addition a large wrap-round touch sensitive bumper is usually fitted. Any contact results in the application of brakes for an emergency stop.

The sonar also serves the purpose of keeping a minimum vehicle separation. An alternative is to divide the track into blocks like a railway and to allow only one truck at a time into any block. Blocking may be used in any case at converging junctions to ensure that two vehicles cannot try to cross the 'points' simultaneously.

MISCELLANEOUS FEATURES

An AGV network usually includes a loop or siding where the trucks can recharge their batteries, automatically making contact with a power supply.

The vehicles usually allow manual driving both for emergency use and so that they can be driven off the network.

AGVs can climb reasonable slopes, but it is also possible for them to use lifts. This requires a length of inductive cable in the floor of the lift, accurately aligned with that outside; appropriate controls and interlocks must be provided so that the AGV can call and enter the lift.

AUTOMATED GUIDED VEHICLES WITH MECHANICAL AND OPTICAL GUIDANCE

Some AGVs use optical sensors to follow a line painted on the floor. It may be painted in fluorescent dye so that it is invisible except when illuminated by ultraviolet lamps on the truck. The main advantage of the painted-line method is that it does not involve cutting grooves and laying cables. Its disadvantages are that the painted line may wear away or become covered over, and that since the track is completely passive it cannot be used for control and communications functions such as switching sections off or modulating a carrier.

Another system is mechanical guiding by a slot in the floor. The slot may even contain a moving chain which each truck can engage for propulsion. This system requires a relatively elaborate fixed installation and is on the borderline between AGVs and fixed conveyors.

FREE-RANGING AUTOMATED GUIDED VEHICLES

Free-ranging AGVs are still in the research stage. Various degrees of autonomy are possible. The most straightforward extension from wire-guided AGVs, and the one which is most highly developed at present, is to set up a network of fixed routes which the trucks follow by dead reckoning (odometry). The network is defined in software rather than being laid down physically. An example of such a system is that being developed at Imperial College, in which the position and orientation of the AGV are continuously calculated by an on-board computer from the distance travelled by each wheel. The truck has two driving wheels with differential steering, and a castor at each corner. Each driving wheel (or a separate measuring wheel) has a incremental encoder of 1000 pulses per revolution which allows circumferential movement to be measured to 0.1 mm. The AGV travels a network made up of straight lines connected by circular arcs of a fixed radius.

Because all odometric navigation systems gradually acumulate errors a secondary navigation system is needed to re-initialize the odometry at intervals. If the journeys are all short this can be done by accurate docking with each workstation. Otherwise some other system must be used. One method is triangulation using optical or acoustic beacons. Another is for the vehicle to detect guide 'marks' when it crosses them; these may be light beams or lines painted on the floor.

The advantage of free-ranging AGVs is that a dense network of routes can be laid down, with many alternative paths so that obstacles (including other AGVs) can be bypassed. Also, of course, it avoids the need to cut grooves in the floor. Changing network layout is purely a software change.

This flexibility implies that scheduling or route planning of the fleet is central to the operation of the system, with frequent reprogramming of the on-board computers being routine. This needs almost constant communication with a central scheduling computer, which is a weakness of the free-ranging AGV concept, since it is not easy to guarantee good communications in the factory environment with its high levels of radio noise and physical obstacles to line-of-sight communication. Some combination of radio and infrared may be necessary. The problem can be avoided, at the cost of losing some of the advantage of free ranging, by not allowing reprogramming of the on-board computers between loading stations.

Bibliographic notes

The best source of information on AGVs is the series of conference proceedings of which the latest is Andersson (1985). A textbook of AGVs which is useful, particularly on wire guidance methods, but not as lucid as might be wished for in an introduction, is Müller (1983).

Chapter 11

Robotics and Artificial Intelligence

As remarked in Chapter 1, AI is a subject with ill-defined boundaries. Its name (coined in 1956 by John McCarthy) has never seemed entirely satisfactory and there are variants such as 'machine intelligence', but nobody has been able to think of a universally acceptable alternative and so the original term has stuck.

AI was originally thought of as a way of formulating and testing psychological theories: an exploration of how thinking might work in principle, whether implemented in a biological brain or not. In philosophy the idea that at least some aspects of thought follow a logical, factually describable process is not new, but only with the advent of computers did it become possible to attempt a mechanical realization of the 'laws of thought'. Logical, but previously exclusively human, skills such as chess playing then became susceptible to mechanization.

This view of AI as a sort of mechanical psychology is still held and, in my opinion, is where its greatest importance lies, but as far as its present relevance to robotics and other practical subjects is concerned AI is just a bag of programming methods. What these methods have in common is that they search for a satisfactory interpretation of data, or a plan of action, among a collection of possibilities, usually on the basis of imperfect knowledge. AI is about search and representation. Representation is the issue typified by questions such as how a model of an object can be stored in a computer in a way which allows effective comparison with an image. This chapter describes some applications of AI to robotics.

Vision

All computer vision systems, even in the relatively simple case of locating an object in a two-dimensional binary image, use methods developed in research in AI. Examples of such methods are algorithms for detecting edges in images, for dividing an image into connected regions, for joining fragmentary edges into long edges, and for fitting models to line data. Such methods of image analysis are now well established and are no longer regarded as within the domain of AI, except when exceptionally advanced. As mentioned in Chapter 5, some of the more interesting sensory tasks such as interpreting an image of a heap of castings cannot be done at present, except in special cases. Work on difficult problems of this kind counts as AI.

Voice communications

There are several situations in which the ability to obey verbal commands would be useful in a robot. Some examples are as follows:

1) teaching a robot by leading through,
2) driving a robot vehicle,
3) controlling a teleoperator,
4) giving a list of required items and their destinations to an AGV or automatic warehouse,
5) emergency override of almost any machine.

Speech recognition systems can be divided into those which recognize isolated words separated by an audible pause and those aiming to understand complete sentences, in which words commonly run together and are hard to isolate from their neighbours.

Isolated-word systems have a vocabulary of typically 100 words, taught by speaking each word five or ten times. One such system is available as a peripheral to microcomputers such as the Apple II, where it is placed between the keyboard and the main processor. Any spoken word it recognizes is turned into a string of characters from a user-defined table, apparently coming from the keyboard. If the user says 'nine', 'return' or 'up', the character 9, the carriage-return character or the pair of characters U P respectively will appear to have been typed. This is completely general purpose and can be used for spoken programming as well as for commands to a robot.

Such systems match each isolated unit of sound with each word in the vocabulary and output the best match. (The matching may be flexible, allowing some variation in length and other parameters.) They are fairly reliable, at least with a single speaker, but still make enough mistakes for some additional means, such as a pushbutton or a special command to which all unrecognized words default, to have to be provided for telling the system when it has made an incorrect identification. There must also be a means, such as a visual display unit or a speech output device, for the system to tell the user what word it thinks has been spoken.

The recognition of connected speed is still in the research stage. The best-known projects so far have been HARPY and HEARSAY-II conducted on behalf of the United States Defense Advanced Research Projects Agency. Carnegie–Mellon University's HEARSAY-II is the more interesting from an AI point of view. It is an expert system whose control structure allows it to try to find matches between spoken sound and a model at several levels simultaneously, from the shortest acoustic segments up to a whole sentence. At any time during an attempt at interpreting an utterance there will be hypotheses at several levels. (A hypothesis is a possible recognition of a syllable, word, phrase etc.) These are generated by program modules called knowledge sources. To begin with, only hypotheses at the lowest level of speech segments can be generated from measurements of the spectrum of the acoustic signal throughout the utterance. As the interpretation progresses, hypotheses are used as inputs by knowledge sources which generate new hypotheses at higher levels. For example, a particular conjunction of syllables will generate a word hypothesis. There are also downward directed knowledge sources. For example, if a low level entity such as a syllable is ambiguous because of uncertain segmentation, it may be possible to identify it by matching with alternative word hypotheses derived at a higher level from other evidence.

The knowledge sources communicate via a common data area called a blackboard, in which are kept the current hypotheses at all levels.

HEARSAY-II and HARPY can understand connected speech from a vocabulary of 1000 words. This has been described as equivalent to a ten year old child.

Current research in Britain under the Alvey programme is aimed at developing speech recognition for a number of practical purposes, such as generating text from speech. Although this is not directed at robotics as such, its success would make the functions listed at the start of this section more practicable. There are still major problems. A notable one is that most speech recognition systems have to be trained with a single person, or very few people; for many applications they will have to become speaker independent.

Planning

A plan is a sequence of actions, or a route, between an initial state and a goal state. A program is a kind of plan, and the conventional way of programming or teaching an industrial robot is for a person to work out a plan, in the form of a sequence of joint movements or of a continuous path in space, and to teach this to the robot.

It is clear that there are times when the automatic generation of a plan would be useful. As flexible manufacturing ·becomes more completely automatic it becomes necessary to plan machining and assembly operations (and indeed other aspects of the operation of the factory). An autonomous vehicle needs to plan its route. If an error occurs during, say, grasping or assembly it would be useful if a robot could plan a new course of action to reverse the error and carry on. Some actions such as deciding where to place the feet of a walking vehicle or how to grasp an awkward workpiece need a plan whose details cannot be formulated in advance.

It is convenient to divide plans into two kinds, those with discrete steps and those consisting of a continuous path. An individual step in a discrete plan may be a continuous path and may take parameters such as the distance or a sensor test. Plans can be hierarchical, i.e. each step in the plan can be made up of smaller steps which themselves need planning.

Planning is part of problem solving, and as such has been one of the central concerns of AI since the 1950s. It has often been studied in the context of game playing, but planning for robots was an obvious field for research as soon as computer-controlled robots were conceivable. Research on problem solving has often taken as its domain the world of a (sometimes simulated) robot building towers out of toy blocks; another research area has been planning the actions of mobile robots.

DISCRETE STEP PLANNING

The basic principle of discrete step planning is ilustrated in Figure 11.1. A robot has to make a valve subassembly by fitting a spring, a washer and a circlip to a valve stem. A plan has to be devised using only the operations 'add washer', 'add circlip' and 'add spring'. The problem can be described in terms of a network (a graph) of arcs representing operations and nodes that represent possible states of the subassembly. In the example the network takes the form of a tree whose root is the initial state. At each node there are three branches representing the three possible operations. A chain of operations forms a path through the tree. Each path is a possible plan, and finding the correct one is done by searching the tree until a path is found which leads from the root (the initial state) to the goal state. As can be seen from Figure 11.1, a plan resulting in a correct assembly is to do the following steps in the order stated:

1) add spring,
2) add washer,
3) add circlip.

For many real planning problems the number of possible plans is too great for such a simple-minded exhaustive search to be effective, and more complex methods such as trying the most likely operations first must be used. Even an exhaustive search can be done in several ways. A planning program can be cast in the form of an expert system if enough is known about the operations and their effects.

ROUTE PLANNING

Planning the route of a mobile robot in a factory or a forest has much in common with finding a collision-free path for a manipulator in a cluttered environment. The mobile robot problem is easier because it can to a large

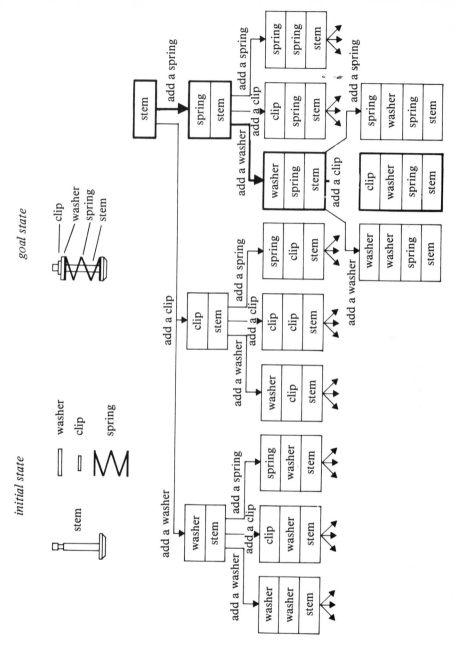

Figure 11.1 *An example of assembly planning. The aim is given the operations 'add a washer', 'add a clip' and 'add a spring' to devise a plan for getting from the initial state to the goal state, in this case a valve assembly in which a spring is held on a stem by a washer and a circlip or split pin. (a) Initial and goal state; (b) the tree of possible assembly plans (only part of the tree is shown) — the correct plan is shown as bold arrows.*

extent be treated as two dimensional (for a land robot, that is; clearly a submersible or a structure-climbing robot can have three-dimensional planning problems).

Although a route is continuous, planning is not always completely different from discrete step planning. Often there is a fixed network of possible paths (arcs) connecting junctions (nodes), and if a measure of the cost of traversing each arc, e.g. its length, is known then planning a route is equivalent to searching for an optimal chain of arcs from the initial node to the goal node. Even if the robot's world is continuous, it can sometimes be modelled by a network. An example is a building in which each room is treated as a network node (Figure 11.2). Within each room other methods have to be used.

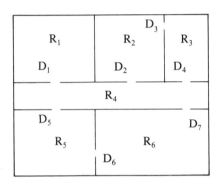 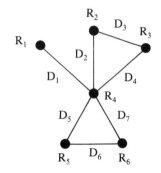

Figure 11.2 *A network representation of rooms and doors.*

Much of the current research on route planning deals with a two-dimensional problem, as illustrated in Figure 11.3, which shows the shortest route for a dustbin-shaped robot in a room cluttered with box-like obstacles. One reason for the popularity of this rather artificial problem domain is that robot sensors, usually sonar, have difficulty making sense of real obstacles such as table legs and electric cables.

Which method can be used depends on whether the robot has a complete model (map) of the problem area or must build it up as it goes along; in the second case trial and error are inevitable, and the robot is unlikely to find the optimum route on the first attempt.

If a map is available and the obstacles can be approximated by polygons, several methods can be used. For example, in Figure 11.3(b) all vertices, including the robot, have been connected (the broken lines). The resulting graph, whose arcs are obstacle edges and vertex-connecting lines, contains the shortest path, which can be found by search. A likely candidate for the shortest path is shown as a heavy broken line. The robot clearly follows walls when it is not crossing the space between vertices.

Figure 11.3(b) treats the robot as a point. One way of taking its finite size into account is to expand all the obstacles by the robot's radius, as

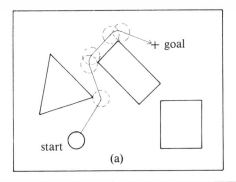

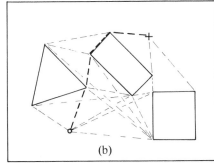
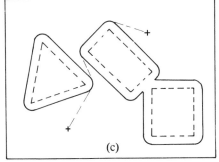

Figure 11.3 *Route planning for a cylindrical mobile robot in a room with polygonal obstacles. (a) The shortest path from the start to the goal. The broken circles show the position of the robot as it approaches and leaves each corner. (b) A graph of arcs connecting all vertices which can see each other, and the robot and goal (treated as points). (c) Obstacles expanded by the robot's radius.*

in Figure 11.3(c), and regard the robot as a point again so that the graph-search method or something similar can be used.

Other methods resemble those a person might use: quasi-analogue, with some trial and error. The program in effect simulates some process which uses elastic strings or propagating waves to find the shortest route. This is computationally expensive; also, any theoretical method of route planning cannot take into account errors in the model. Therefore, for a practical robot it may often be better to do as little planning as possible. Instead, previously explored routes can be used, perhaps being modified by planning if not quite right for a new journey. Another method of finding a route is trial and error or 'blundering': a good example of the utility of this is when the goal can be seen through an open forest of randomly located and unmapped obstacles, such as a literal forest. Often only a combination of these methods will be successful.

Modelling

Underlying several robot functions such as vision and planning is the need to *model* the environment or part of it, i.e. to find a way of describing objects and spatial relationships in some language or *representation*. This is also a requirement of computer graphics and CAD, and similar techniques are sometimes used. A basic approach is to represent objects by polyhedra or combinations of cylinders, whose surfaces can be defined by the methods of coordinate geometry. In robotics it may be necessary to establish relationships such as attachment between objects so that whatever changes are applied to one also affect the other; an example is the relationship between a gripper and the object it holds. Representation of this kind, allowing attachment, is provided in some robot programming languages such as RAPT.

Manipulation robots need three-dimensional modelling; robot vehicles operate on a surface and what must be represented is maps and related information. (Solid modelling may still be needed as well.) An example of two-dimensional modelling is the representation in Figure 11.2.

Adaptive control

This term is used in servomechanism theory for control systems in which parameters such as amplifier gains are modified by feedback. In the present context it means changing or modifying a series of actions in the light of sensor data. It can make use of planning: the sensor data is used to determine the initial state and then a planner is called to devise a series of actions leading to a goal.

An example of adaptive control is the grasping of an object of unknown shape or orientation by a hand having a range of actions available such as probing with one finger or cupping several fingers round the object. Adaptive control in this case means using the information gathered at each stage to select the next action until a secure grasp is achieved and the orientation of the object is known.

Sensor-based operations such as vision-controlled welding or the use of force sensing during assembly are sometimes described as adaptive control.

Error monitoring and recovery

Error recovery is a potentially important example of adaptive control. A work cell consisting of machine tools, robots, feeders and conveyors is subject to problems such as tools that break, feeders that jam and incorrect components supplied. The aim of automatic error recovery is to monitor the operation of the cell with sensors and to take corrective action if a fault occurs. This implies the need to work out what has happened, or what the current state is, and then either to call some pre-programmed procedure or to plan a new course of action. If the fault is beyond the system's ability to correct, perhaps because of a major mechanical failure, the aim is to provide the maintenance staff with a diagnosis of the fault and its cause.

An error monitoring and recovery system called AFFIRM is being developed at the University College of Wales, Aberystwyth. It is a knowledge based system which can give a detailed explanation in English of the sequence of events preceding an error. The focus of research is as much on understanding faults as on correcting them. It uses the *frame* formalism invented by Minsky at MIT. A frame is a structure holding knowledge about a situation in such a way that expected values and events can be held. In AFFIRM a task such as moving an object is represented in a frame, with links to its component actions and eventually to elementary actions and sensor tests. The successful completion of an action is signalled by a matching of actual and expected sensor values. A record is kept of the actions and tests, and if the system halts this can be used for diagnostic processing. Recovery is envisaged as taking some simple action such as rejecting a part, or selecting an error recovery program from a library.

In principle it is possible for a planner to generate a corrective action plan. The utility of this is likely to be limited by the robot's inability to repair damaged equipment. At the lowest level such as failure to grasp a part properly, error recovery planning merges with adaptive control.

Autonomy and intelligence in robots

Autonomous robots are usually taken to mean free-ranging mobile robots which are not teleoperated but plan and execute their own actions. Autonomy is rarely complete, since robots generally work on goals set by human users and there is often a degree of supervisory remote control, but a robot may be unsupervised for the duration of some task which it has to carry out using only its own resources.

The term 'intelligence' as applied to robots usually means the ability to do rather limited processing of visual images, to use sensory feedback or to do simple planning. Such abilities are sufficient for many industrial tasks and many, perhaps most, robots will never be more intelligent than this. However, mobile robots (and to some extent industrial robots doing complex work such as assembly) will need much more advanced abilities if they are to be fully effective in complex environments; some of these abilities such as navigation and route planning have been touched on. Further, there is the subject of robots as tools for AI research, mentioned in Chapter 9. Even here most of the 'intelligence' being studied is in narrow domains such as assembly planning; but when we consider how to make a mobile robot function autonomously in a natural environment the intelligence it must have starts to have more in common with intelligence as manifested in people and animals.

The problem of designing an intelligent system can be expressed as the problem of designing a control structure. (There is an implicit assumption here that intelligence can be meaningfully interpreted in this way. As long as attention is confined to practical problems of robotics it is probably true, but whether this approach can ever encompass the more subjective aspects of mentality such as emotion and consciousness is an open question.) Whether the goal is to build effective robots or to understand intelligence, the design of control structures for autonomous robots (or the modelling of human or animal control structures) is a subject in its own right. A control structure in this sense is the most global level of organization of the robot's (or the brain's) software. Such structures or architectures are often represented by block diagrams such as Figure 11.4. Such diagrams occur in robotics, ethology and psychology.

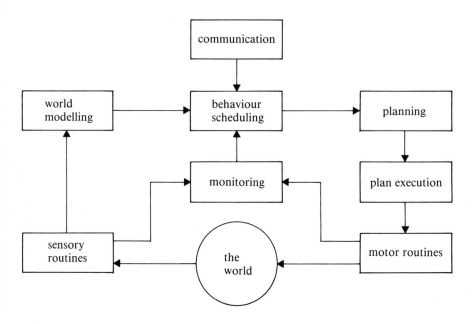

Figure 11.4 *A block diagram representing a control structure as the most global level of organization of the robot's software.*

Designing a control system for an 'intelligent' or 'autonomous' robot is a matter of deciding on such a block diagram and of designing what goes in the boxes, and what information flows between them.

Some of the boxes contain subsystems for functions such as modelling, vision, route planning and navigation which have already been discussed. In addition there may need to be modules for deciding what task to do next (the 'behaviour scheduler' of Figure 11.4), for monitoring the system's actions and for detecting important events.

There is no universal agreement on what constitutes intelligence or what functions are needed. Some of the issues are listed below.

First, it is possible to be autonomous and stupid (like lower animals, which cannot be switched off or reprogrammed but have only limited intelligence) or intelligent but with little autonomy; but it seems to the author to be almost a defining characteristic of the most intelligent systems that they would not allow themselves to be turned off or reprogrammed. Also, Nilsson has suggested that an intelligent system, a 'computer individual', needs to exist for an indefinite period, learning all the time as a person does, rather than being switched on each day like most machines.

A related idea is that an intelligent system's world model should allow it to keep track of its situation at all times. A person can almost always do this: you may get lost, but you retain a coherent model of your relationship to your surroundings and continue to act appropriately. The opposite of this is a robot which jams against an obstacle without any of its sensors detecting the collision, and continues to grind away indefinitely. Put another way, an essential component of intelligence is error recovery.

Another issue in autonomy is the degree to which a system sets its own goals. Even animals and people are set goals not of their own making. The question of how far an autonomous system must set its own goals has no simple answer.

These issues are all at an early stage of exploration. Indeed, since they merge into the psychological and philosophical end of AI there is no reason to suppose that they will have a neat resolution. Rather, there will be a never ending quest to make robot intelligence more and more like the real thing.

So far, although many robots have been built which conform roughly to the structure of Figure 11.4, their implementation of most of the modules in it has been rudimentary and there has been no attempt to exploit the idea of the continuing, learning individual with a never-failing world model. Perhaps the most complete intelligent robot so far was Shakey (Chapter 9), which was fairly advanced in sensing, modelling, planning and execution (although it has been surpassed in some of these areas) and only lacked autonomy of scheduling behaviour to be very much like an animal.

Expert systems in robotics

Expert systems, sometimes known as intelligent knowledge based systems, have been an area of spectacular growth in recent years; indeed, some of those promoting expert systems tend to regard them as superseding old-fashioned AI. This view is unlikely to stand the test of time, in the author's opinion; rather, the knowledge based way of looking at problems will be seen as an important approach to computerized problem solving but no more than that.

The basic principle of expert systems is that, when writing programs to solve problems such as diagnosis, a logical separation can be made between *knowledge* about the problem and the (software) *machinery* needed to operate on the knowledge, together with a statement of the problem, to find a solution. The knowledge in such a system can be expressed in a uniform way, often in the form of statements or rules relating, say, symptoms to possible causes or to intermediate conclusions which in turn are referred to by other rules. The application of a rule to yield a conclusion given a fact is called inference, and it is possible to form a logical chain of inferences connecting the statement of the problem to an ultimate conclusion. The procedure for doing this is handled by a part of the program often called an inference engine; the expert system then consists of the inference engine together with the rules.

A characteristic of many expert systems is a sophisticated user interface which can present complex information in an easily understandable way. For example, a diagnosis would make clear the line of reasoning leading to it. The system might also suggest a course of treatment.

Some of their applications to robotics are indirect in that expert systems may be used in subsystems such as vision or voice input (e.g. HEARSAY). Of more specifically robotic functions, some such as assembly planning and error recovery can be approached using expert systems; an example is AFFIRM, discussed earlier. The Metatorch arc welding guidance system also uses knowledge based methods. Many other uses will soon emerge.

Bibliographic notes

There are dozens of texts on artificial intelligence; among the clearest of those recently published is Charniak and McDermott (1985). The textbooks are useful for background but usually of limited relevance to robotics. The best source is once again conference proceedings such as the *Robot Vision and Sensory Controls* series (e.g. Pugh (1984)). Other series are the *International Joint Conferences on Artificial Intelligence* (IJCAI) and the *American Association for Artificial Intelligence* (AAAI).

A recent conference on speech recognition and synthesis is Holmes (1984).

Among the many books on expert systems, two straighforward introductory texts are Addis (1985), and Alty and Coombs (1984), although in common with most books on expert systems they contain little of direct relevance to robotics.

Chapter 12
Economic and Social Aspects of Robotics

Reasons for installing robots

Industrial robots (programmable manipulators) can be regarded as replacements for human workers. A list of reasons for replacing people by robots is given shortly. They can also be installed as part of a flexible automation system such as a manufacturing cell; in this case it is dedicated automatic machinery, rather than people, with which robots compete. Similarly an AGV can replace a man with a fork lift, or a fleet of AGVs may replace a conveyor-style assembly line. A teleoperator does not replace a person but extends his capabilities.

Some reasons for installing robots are as follows:

1) to save money, if the cost of a robot over its lifetime is less than the cost of employing a person (or persons if it replaces more than one),
2) to increase the speed of an operation and so increase production,
3) to improve product quality through improved consistency,
4) to handle loads too heavy for a person,
5) to handle dangerous loads (radioactive, explosive, toxic),
6) to eliminate boring or unpleasant work,
7) when a person cannot gain access, e.g. in narrow tubes,
8) when a task needs movements too precise for a person to make.

The economic reasons are discussed in the next section. They are the main factor responsible for the growth of the robot industry. Nevertheless the benefits in human terms should not be underestimated. Many industrial tasks are unpleasant or dangerous and well worth eliminating. In the case of jobs which are boring rather than dangerous the benefit of eliminating this work has to be weighed against the effect on the displaced worker: he or she may transfer to more interesting work, or equally boring work or be made redundant.

Economic costs and benefits of installing industrial robots

From the list of reasons just given, we can see that there are potential benefits of installing robots, but in any particular situation the decision whether to use robots or not must be justified in detail. In many cases the decision is primarily an economic one, and it is possible to quantify the costs and benefits, as outlined shortly; in others there may be no alternative to robots, so they must be used whatever the cost. This is normal with teleoperators, but may sometimes apply to industrial robots as well.

The economic costs and savings of replacing a man or men by a robot are these.

Costs:
   1) purchase price,
   2) special tooling,
   3) installation,
   4) staff training,
   5) maintenance,
   6) power,
   7) finance,
   8) depreciation.

Savings:
   1) labour displaced,
   2) quality improvement,
   3) increase in throughput (not always positive),
   4) savings on quality of working conditions.

If these items can be quantified it is possible to calculate a payback period. If a robot is considered as a replacement for a person it is possible to calculate how long the robot must work before it shows a net saving. This is the *payback period,* and is given by:

$$P = \frac{I}{L - E}$$

where P is the payback period in years, I is the initial cost, L is the total annual labour costs replaced by the robot and E is the annual maintenance cost. The payback period must obviously be less than the life of the robot, which might be five or ten years, and in fact payback times of three years or less are sought. Clearly the payback period is low when wages are high and vice versa, so in Britain where wages are lower than in some other industrial countries there has been less of an economic incentive to install robots and other forms of automation.

The payback period is reduced if the robot replaces more than one worker, usually by working two or three shifts.

The formula given here is the simplest of several. More complicated calculations can take into account the change in production if the robot operates its associated machinery faster or slower than the man it replaces, and there are financial methods such as the *annual return on interest* calculation and methods of taking into account interest rates and inflation.

The list of costs and savings and the method of payback analysis assume that a robot does a job which can be done by human workers, so a straightforward comparison is possible. Increasingly, however, robots are designed into manufacturing cells which could probably not work efficiently at all with human labour. In this case the cost comparison should be between the robot-based cell and a dedicated machine for doing the same task. Theoretically, almost any manufacturing process can be carried out by (a) human workers operating individual machines, (b) robots loading individual machines or (c) a dedicated machine doing the whole process. Mixtures of these methods often occur. It is usually argued that, as the volume of production rises, the most economical method progresses from (a) through (b) to (c). Other considerations are the belief that labour costs will rise faster than those of automation, and the possibility that in future many products will be made in such a variety of models and in production runs so short that the flexibility of robot-based automation will outweigh the advantage of dedicated machines.

Reliability

A measure of reliability is *uptime*: the percentage of the time for which the robot is available. It is equal to 100% minus the *downtime*, which is the percentage of the time for which the robot is out of service for repair or maintenance. An uptime of 97% or 98% can be achieved, which is enough for many industrial processes but not all, and so in critical cases some form of back-up must be provided. The *mean time between failures* (MTBF) is self-explanatory, except that it should be noted that it refers to the working time, not calendar time. Values of several hundred hours are typical. The *mean time to repair* (MTTR) is the average duration of a repair. In some situations any major fault may cause the robot to be shut down for the rest of the shift, or until a repair team is available, and so the MTTR may well be much longer than the actual repair time.

These quantities are related:

$$\text{downtime} = \frac{\text{MTTR}}{\text{MTBF}}$$

For example, if the MTBF is 400 h and the MTTR is 8 h, the downtime is 2% and the uptime is 98%.

Safety and environmental factors

As well as the need to guarantee the safety of people in the workplace, there are several other safety issues in robotics. Some of those included here are arguably questions of reliability:

1) safety of human beings where robots are working,
2) safety benefits due to robots,
3) prevention of mechanical damage by robots to other equipment,
4) prevention of damage to the robot,
5) avoidance of fires and pollution produced by robots,
6) protection of robots from adverse environments.

SAFETY OF HUMAN BEINGS

The first accidents involving industrial robots, including at least one fatality, have already happened. The unique danger of robots lies in their unpredictability, in part because the variable configuration makes it hard to tell where the arm can reach, but mainly because there is no way to tell just by looking at a robot what it is going to do next. If it is repeating a cycle, it is likely to go on doing it, but this is not certain; it could be about to reach the end of a fixed number of operations before doing something else. The greatest danger, because there is no pattern of movements to observe, is when the robot is stationary. This could be because it is waiting for some event, such as an interlock signal, so that it may suddenly spring to life. Many robots can move a heavy payload with great force at speeds of 1 ms^{-1} or more and can easily kill or seriously injure.

The obvious precaution is to cage a robot in, but there is always some need to go into the cage, e.g. to service the machinery within it, and it may be necessary to do this with the power on so that the robot can be moved. In this case it is almost impossible to guarantee safety absolutely.

An extension of the cage concept is to use light beams and pressure mats to detect the presence of intruders into the danger area. A more sophisticated approach is to fit the robot with proximity sensors on its moving parts and to couple these to the controlling program so that if the robot detects an impending collision it stops. This is difficult to do comprehensively, particularly if it grasps large workpieces which cannot be guarded in this way. A more advanced extension of this idea is to fit the work cell with a vision or range imaging system which would track a person and stop the robot if necessary. These active sensing systems have the disadvantage of being complex and not necessarily fail-safe.

As with any machine, emergency stop buttons can be provided in addition to active or passive guards. The effect of such buttons should be carefully considered; simply cutting power off may cause the robot's load to sag under its own weight, or trap a person with no safe way of freeing him.

AGVs cannot, of course, be fenced in. Their main safety precautions are (a) low speed, (b) clearly marked routes, (c) turn indicators, horns and flashing lights, and (d) touch and proximity sensors which can stop the vehicle quickly.

A danger with legged robots is of their becoming unstable and falling over, endangering the crew if any and anyone nearby. A second danger is that of treading on people, as the author can testify from personal experience. Little attention has been paid to these problems. (In Figure 9.18 the GE quadruped has clearly been fitted with tubular metal outriggers to prevent it falling over too far.)

SAFETY BENEFITS OF USING ROBOTS

Despite their dangers, on balance robots can improve safety by saving people from dangerous tasks such as reaching into presses and injection moulding machines, lifting heavy loads and working in dangerous environments.

PREVENTION OF MECHANICAL DAMAGE

The problem is essentially the same as that of ensuring human safety, but is easier since most equipment is static and if it moves it does so predictably. One precaution which can be taken is to limit joint torque by friction clutches, by limited motor power or by sensing loads and cutting power off if some limit is reached.

AVOIDANCE OF ROBOT-GENERATED FIRES AND POLLUTION

The problems are not unique to robots. Any hydraulic oil which leaks is a fire hazard, and may also be unacceptable if high cleanliness is desired, as in food processing. Electric joint motors can produce sparks from commutators or faulty connections and so may be unacceptable in explosive atmospheres; other emissions such as heat, vibration and acoustic and electrical noise may have to be controlled.

PROTECTION OF ROBOTS FROM ADVERSE ENVIRONMENTS

Among the damaging influences to which robots may be exposed are the following:

1) radiation (nuclear applications),
2) extremes of temperature (foundries),
3) abrasive particles (grinding),
4) sparks and molten metal splashes (welding, casting),
5) clogging particles (paint spraying),
6) corrosive chemicals (investment casting),
7) shock and vibration (forging),

8) electrical noise (any factory),
9) water and other liquids (from coolant sprays and washing),
10) steam (from steam cleaning).

The precautions to be taken are the same as for any machine: flexible coverings on joints, circulating filtered air at positive pressure throughout the robot, non-flammable and corrosion-resistant materials, mains filters or uninterruptable power supplies and so on.

Acceptability of industrial robots by the workforce

Whether the introduction of robots is welcomed or resisted by the workforce of a particular factory depends on several factors. These may not be of equal importance to all those involved: the official union view, the view of a particular worker and that of the various layers of management may differ.

Engelberger quotes several instances of a positive response to robots among factory workers, and it is not always correct to assume that the management will be in favour of robots and the shop floor opposed to them. In Britain the engineering trade unions have given a cautious welcome to robotics, perhaps believing that the choice is not between robots and jobs but between robotized industry with at least some jobs and no industry at all.

The likelihood of robots being welcomed is increased if the following conditions are met:

1) they do not lead directly to redundancies, but the displaced workers are transferred to other work;
2) they are built into a new installation such that there is no traditional manning level for comparison (in this case the robot is merely part of a larger piece of automation being introduced and does not directly replace anyone);
3) the tasks they take over are unpleasant or dangerous;
4) they result in the workers being retrained to do more skilled and interesting jobs, e.g. overseeing a robot-operated process and reprogramming the robots (of course, the new job may be less interesting than the old one);
5) they are perceived as enhancing the status of those who work with them;
6) they create new jobs such as programming and maintenance;
7) as with any new equipment, the ease of introduction depends on the state of industrial relations in the factory.

Employment

Even though robots can often be introduced to a factory without an immediate loss of jobs, their main rationale is after all to save labour costs and so over the years there must be a loss of employment, mainly jobs which would have been created in the absence of automation. The scale of this is not agreed. Some argue that so far the number of robots is tiny compared with that of a country's workforce, and also that jobs are created in robot manufacture, installation and maintenance which compensate to some extent for the loss of employment in manufacturing.

It is probably pointless to try to work out the impact of robots in isolation; they are merely an aspect of automation. There is no doubt that in industrial countries competitive pressure will lead to a steady increase in automation and, assuming that total production does not also increase steadily, the implication is that fewer and fewer hours will need to be worked in manufacturing. The present causes of high unemployment may not have much to do with automation, but it is unlikely that any future manufacturing industry will be as labour intensive as those of the past.

Other social issues of robotics

This section suggests some questions which can be asked about the social implications of robots (including mobile robots and teleoperators). It does not in general attempt to answer them, but comments on some, where relevant facts are available. One or two have been discussed already.

SOCIAL ISSUES

1) Should the use of industrial robots be encouraged or opposed?
2) Is an increasing use of industrial robots inevitable?
3) When they are introduced, what safeguards should be established?
4) Is there a real prospect of increased military use of robotics, and what are the consequences?
5) Do robots for police and security work pose a threat to freedom?
6) Does robotics have any real potential to enhance people's lives by, for example, relieving them of drudgery and dangerous work?
7) Are domestic robots feasible, and if so are they desirable?
8) Is robotics likely to enhance or diminish economic equality, both within and between nations?

Some of the questions are simplistic and prompt further questions. For example, the first invites questions such as 'encouraged by whom?' and 'according to what criteria should the benefits be assessed?'. All these issues have moral and political ramifications, and tend not to have simple value-free answers. They are listed mainly to emphasize the fact that there is more to the social dimension of robotics than its effect on employment. Some specific issues will now be discussed.

MILITARY ROBOTICS

So far, the main form of military robotics (apart from guided missiles and RPVs) has been the use of teleoperated vehicles for bomb disposal and allied tasks, on a very small scale. Robotics may soon allow legged vehicles to be built, but this would not have much effect on the character of warfare.

A more radical step would be the introduction of autonomous robot vehicles for battlefield and support use, as described in Chapter 9. Such a development can be seen as good in that it removes men from danger, or as bad in that it might lead to yet another expensive dimension of escalation for military technology.

POLICE AND SECURITY USES OF MOBILE ROBOTS

Mobile robots are being developed for use as sentries. They could also play the role of a nightwatchman, or a guard dog let loose in, say, a warehouse. The temptation to arm such robots with guns and other weapons is likely to prove irresistible to some, and this will inevitably lead to accidental shootings (accidental, that is, from the point of view of the robot's owners; at present, the robot would be unlikely to be conceded a point of view). The most practical legal and moral way of looking at this problem is perhaps to consider the robot as equivalent to a dangerous animal.

Mobile robots, for observation only or equipped with weapons, may also come to be used by police in controlling or fighting crowds.

TELEOPERATORS

Teleoperators do not replace people but allow them to do in safety jobs which would otherwise be dangerous or impossible. If their use has an aspect which is cause for concern it is that teleoperators make it easier to develop and manufacture nuclear, chemical and biological weapons.

MEDICAL USES OF ROBOTICS

The use of telemanipulators by the disabled was discussed in Chapter 8. This may be extended to mobile robots for fetching things in the home, although there are obstacles to this becoming practicable soon. The other main category of medical robotics is mobility aids and devices: prosthetic and orthotic legs and walking chairs. None of these has been very satisfactory so far but progress seems likely.

Bibliographic notes

Many books such as Engelberger (1980) and Owen (1985) discuss economics, reliability and safety aspects of robotics. A relevant conference series is *Human Factors in Manufacturing*; see for example Bullinger (1985). Safety, training and economics are also discussed in *Robotics Trends: Applications, Research, Education and Safety (Proceedings of the 8th Annual Conference of the British Robot Association,* IFS Publications, 1985).

A textbook on robot safety is Bonney and Yong (1985).

References and Bibliography

These references are intended as suggestions for further reading to broaden and deepen knowledge of the subjects introduced in this book, and not as a comprehensive guide to the latest research. Therefore they refer, on the whole, to a few books rather than to many papers.

Notes and Abbreviations

A second year in square brackets is the year of the conference if this was earlier than the year of publication. Abbreviations of conference series and journal titles are as follows:

AAAI: American Assocation for Artificial Intelligence
CISM: Centre International des Sciences Mécaniques
ECAI: European Conference on Artificial Intelligence
ICAR: International Conference on Advanced Robotics
IFAC: International Federation on Automatic Control
IFToMM: International Federation for the Theory of Machines
            and Mechanisms
IJCAI: International Joint Conference on Artificial Intelligence
IJRR: International Journal of Robotics Research
ISIR: International Symposium on Industrial Robots
ISTVS: International Society for Terrain Vehicle Systems
PWN: Polish Scientific Publishers, Warsaw
RoViSeC: Robot Vision and Sensory Controls
RoManSy: Theory and Practice of Robots and Manipulators (Proceedings of the
            CISM-IFToMM Symposia, a biennial series)

Most of the conferences listed above are annual or biennial. The number of journals devoted to robotics is growing. IJRR is, on the whole, at the mathematical end of the spectrum; *Industrial Robot* reports practical and commercial developments. A new journal on robot applications is *Robotics*; other journals are *Robotica* and the *Journal of Robotic Systems*. A popular but good magazine is *Robotics Age*. Some journals such as *Digital Systems for Industrial Automation* carry a lot of robotics papers. Organizations such as the Institute of Electrical and Electronics Engineers and the ACM also produce a variety of journals relevant to robotics and artificial intelligence. There is also a journal called *Artificial Intelligence*, and several devoted to image analysis and pattern recognition, often in the context of subjects such as medicine and remote sensing.

Addis, T.R. *Designing Knowledge-based Systems* Kogan Page, London, 1985.
Alloca, J.A.; Stuart, A. *Transducers: Theory and Applications* Reston Publishing Co. Inc., 1984.
Alty, J.L.; Coombs, M.J. *Expert Systems* NCC Publications, 1984.
Andersson, S.E. (ed) *Automated Guided Vehicle Systems 3* (Proceedings of the 3rd International Conference, Stockholm 1985) North-Holland, 1985.

Bardelli, R.; Dario, P.; DeRossi, D.; Pinotti, P.C. Piezo- and pyroelectric polymers: skin-like tactile sensors for robots and prostheses. In *13th ISIR* April 1983.

Binford, T.O. Survey of model-based image analysis systems. *IJRR* Spring 1982, 1 (1), 18-64.

Bonney, M.; Yong, Y.F. (eds) *Robot Safety* IFS Publications, 1985.

Brady, M.; Hollerbach, J.M.; Johnson, T.L.; Lozano-Perez, T.; Mason, M.T. *Robot Motion: Planning and Control* MIT Press, 1983.

Bullinger, H.J. (ed) *Human Factors in Manufacturing* (Proceedings of the 2nd International Conference, Stuttgart 1985). IFS Publications, 1985.

Cherniak, E.; McDermott, D. *Introduction to Artificial Intelligence* Addison-Wesley, 1985

Clocksin, W.F.; Bromley, J.S.E.; Davey, P.G.; Vidler, A.R; Morgan, C.G. An implementation of model-based visual feedback for robot arc welding of thin sheet steel. *IJRR* 4, (1), Spring 1985, pp. 13-26.

Coiffet, P. *Modelling and Control* Kogan Page, London, 1983.

Dario, P.; Domenici, C.; Bardelli, R.; De Rossi, D.; Pinotti, P.C. Piezoelectric polymers: new sensor materials for robotic applications. In *13th ISIR* April 1983.

Engelberger, J.F. *Robotics In Practice* Kogan Page, London, 1980.

Gruver, W.A. *et al.* Evaluation of commercially available robot programming languages. In *13th ISIR* Chicago 1983, pp. 12.58-12.68.

Heginbotham, W.B. (ed) *Assembly Automation 1985.* (Proceedings of the 6th international conference, Birmingham 1985) North-Holland.

Hirose, S.; Ikuta, K.; Umetani, Y. A new design method of servoactuators based on the shape memory effect. *RoManSy 84 5th CISM-IFToMM Symp.* Kogan Page, London, 1985, pp. 339-350.

Hirose, S.; Umetani, Y. Kinematic control of an active cord mechanism with tactile sensors. *2nd CISM-IFToMM Symp. on Theory & Practice of Robots & Manipulators* Elsevier, 1977 [1976], pp. 241-252.

Hirose, S.; Umetani, Y.; Oda, S. An active cord mechanism with oblique swivel joints, and its control. *4th CISM-IFToMM Symp. on Theory & Practice of Robots Manipulators* PWN, Warsaw, 1983 [1981], pp. 327-340.

Holmes, J.N. (ed) *Speech Technology.* (Proceedings of the 1st International Conference, Brighton 1984) North-Holland, 1984.

Jacobsen, S.C.; Wood, J.E.; Knutti, D.F.; Biggers, K.B.; Iversen, E.K. The Version 1 Utah/MIT Dextrous Hand. *Robotics Research: the 2nd International Symp.* MIT Press, 1985, pp. 301-308.

Kato, I. Mechanical hands illustrated. *Survey Japan 1982* (distributed by Springer-Verlag and Hemisphere Publishing).

Kobayashi, H. On the articulated hands. *Robotics Research: the 2nd International Symp.* MIT Press, 1985. pp. 293-300.

Larcombe, M.H.E.; Halsall, J.R. *Robotics in Nuclear Engineering* Graham & Trotman, 1984.

Lhote, F.; Kauffmann, J-M.; André, P.; Taillard, J-P. *Robot Components and Systems* Kogan Page, London, 1984.

Lozano-Perez, T. *Robot Programming* Massachusetts Institute of Technology AI Memo number 698, December 1982.

Mason, M.T.; Salisbury, J.K. *Robot hands and the mechanics of manipulation.* MIT Press 1985.

McCloy, D.; Martin, H.R. *Control of Fluid Power* Ellis Horwood, 1980.

Mott, D.H.; Lee, M.H.; Nicholls, H.R. An experimental very high resolution tactile sensor array. *RoViSec 4* London, 1984. IFS Publications, 1984.

Muller, T. *Automated Guided Vehicles* IFS Publications, 1983.

Owen, T. *Assembly with robots* Kogan Page, London, 1985.

Paul, R.P. *Robot Manipulators: Mathematics, Programming and Control* MIT Press, Massachusetts, 1981.

Pugh, A. (ed) *Ro ViSeC 4* (proceedings of the 4th international conference on robot vision and sensory controls. London 1984. IFS Publications, 1984.

Reichardt, J. *Robots: Fact, Fiction & Prediction* Thames & Hudson, 1978.

Rosenfeld, A.; Kak, A.C. *Digital Picture Processing* (2nd edition) Academic Press, 1982.

Russell, R.A. A thermal sensor array to provide tactile feedback for robots. *IJRR* Fall 1985, **14**, (3), pp. 35-39.

Stackhouse, E. A new concept in wrist flexibility. *9th ISIR* Washington D.C., 1979.

Sutherland, I.E. *A Walking Robot* The Marcian Chronicles Inc., P.O. Box 10209, Pittsburgh, PA 15232, 1983.

Tanie, K.; Komoriya, K.; Kaneko, M.; Tachi, S.; Fujikawa, A. A high resolution tactile sensor. *Ro ViSeC 4* IFS Publications, 1984.

Taylor, W.K.; Lavie, D.; East, I.I. A curvilinear snake arm robot with gripper axis fibre optic image processor feedback. In *Robotica* 1, 1983.

Thring, M.W. *Robots and Telechirs* Ellis Horwood, 1983.

Todd, D.J. *Walking Machines: An Introduction to Legged Robots* Kogan Page, London, (Methuen in the USA) 1985.

Vertut, J.; Coiffet, P. *Teleoperations and Robotics: Evolution and Development* Kogan Page, London, 1985.

Vukobratovic, M.; Potkonjak, V. *Dynamics of Manipulation Robots* Springer-Verlag, Berlin, 1982.

Vukobratovic, M.; Stokic, D. *Control of Manipulation Robots* Springer-Verlag, Berlin, 1982.

Index